MG CLASSICS BOOK 2

1945-1965

by Eric Dymock

Produced in Great Britain by Dove Publishing Ltd.

5 Abbey Park, Torksey, Lincoln LN1 2LS

ISBN 978-0-9574585-8-1

This completely revised updated edition is based on The MG File,

an Eric Dymock Motor Book by Dove Publishing,

Old West Kirk Manse, 31 Argyle Terrace, Rothesay, Bute PA20 0BD in 2001.

(ISBN was 0 9534142 3 X)

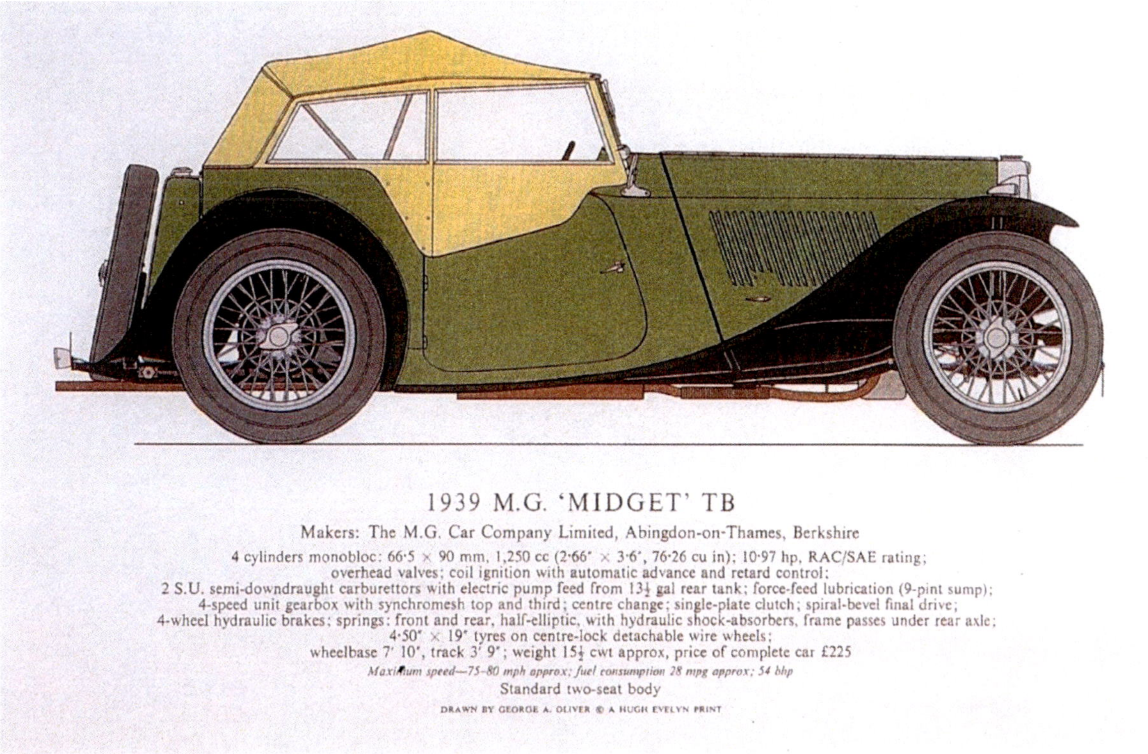

The last pre-war MG. Scottish motor historian George Oliver portrays the 1939 TB Midget for a Hugh Evelyn print. In 1945 MG production restarted pretty much from where it had left off.

Contents

MG Classics History: 1945-1965

Cecil Kimber halted car production at Abingdon as soon as war was declared in September 1939. In its 17 years MG had made some 22,500 cars but, with a national emergency approaching, industry knew at once it would need to adapt for war work. The Munich crisis and years of concern over German aggression had fostered contingency plans that now took effect. MG's stock of car parts and production facilities were stored nearby in Abingdon, while emergency measures were worked out by government. MG secured contracts for overhauling Carden-Lloyd light tanks and later heavier Matildas. A three-month initiative for work on Browning fighter aircraft guns had to be abandoned when it was pointed out that MG had no firing range. A substantial contract for assembling trucks from kits crated in America so filled the factory that extra space was needed. Building materials were becoming difficult to obtain so maintenance staff made an extension of the press shop out of the huge packing cases in which the components arrived.

Two AEC diesels and a Wilson epicyclic gearbox were features of the A12 Infantry tank Mark II Matildas, of which 143 were overhauled in the MG works, once it was equipped with an overhead crane.

There were far-reaching consequences from one project to which Kimber and works manager George Propert signed up. MG took on the construction of the entire forward section of a twin-engined aircraft, the Albemarle AW41. Designers at Armstrong Whitworth had been working on it since 1938 with a brief to meet shortages of raw materials like aluminium in the event of war and the probability, it seemed to the Air Ministry, of the established aircraft industry suffering sustained bomb damage. In that event it might be unable to function and compromise projects had to be in place for concerns like MG, or others unaccustomed to aero engineering, to take on some tasks.

De Havilland worked to a similar mandate for the DH98 Mosquito, ready in November 1940 for its inaugural flight, but the Albemarle was less ambitious. A medium bomber and reconnaissance aircraft the 77ft (23.5m) wingspan Albemarle was made of spruce and unstressed plywood over a tubular stainless steel frame. Its Bristol Hercules 14-cylinder sleeve-valve air-cooled radial engines were good for 300mph (483kph) and a range of 4,000 miles (6400km). The Royal Air Force's first operational nosewheel undercarriage aircraft, Albemarles took part in the invasion of Sicily, D-day in 1944, and towed gliders to Arnhem.

It must be remembered that MG had essentially constructed cars from pieces made elsewhere in the Nuffield group. It did not make components itself so it had no stock of machine tools or presswork capacity. It had to recruit a workforce to make production jigs for assembling the entire front end of the Albemarle's fuselage with the cockpit and the pilot's and navigator's control systems. An elaborate test rig was built to ensure that every one of over 600 made not only worked perfectly but would also match up precisely to the rest of the aircraft that was made elsewhere. More aero-related projects followed, including Lancaster engine mountings, components for Typhoons and wing spars for Hawker Tempests. Work on tanks developed with turrets for Crusaders, and converting Shermans Churchills and Centaurs for special functions. This included 3000 sets of equipment for DD (Duplex Drive) tanks enabling them to wade ashore on D-Day.

The Albemarle contract proved the undoing of Cecil Kimber. His failure to consult Lord Nuffield on its detail was regarded as unforgivable and in 1941 he was sacked. William Miles Webster Thomas (1897-1980), former journalist and now managing director of Morris Motors, brought Kimber the bad news. Sir Miles, also destined to face difficulties with Nuffield, later thought Kimber courageous but warnings for neglecting profit-making had gone unheeded. Harold Alfred Ryder (1888-1950) an executive with a long record of service to Nuffield at Morris was appointed managing director instead.

Following his dismissal Cecil Kimber joined Charlesworth Motor Bodies, which had been making MG bodies since 1936, where he reorganised an aircraft factory through 1942. He then took up an appointment with Specialloid Pistons, which was introducing a new range of split skirt designs, yet with his life's work cut short he was almost inconsolable. He worked conscientiously, although it seemed unlikely he would be part of any sports car revival after the war. On Sunday 4th February 1945, just short of his 57th birthday, he was one of two fatal casualties in a railway accident outside King's Cross. During a business assignment for Specialloid on the LNER service to Leeds, the A4 Pacific "Silver Fox" locomotive spun its driving wheels on a newly laid track. Ascending the 1 in 105 incline inside the middle of the gloomy Gas Works tunnels the driver was apparently unaware his train had stopped and was sliding slowly backwards. A signalman changed the points to divert it on to an empty track but it was too late. A coach overturned, crushing Kimber's first class compartment.

Post-war euphoria in the motor industry was short-lived. Wartime contracts were terminated. There was no cash for new models. Petrol remained on ration throughout the 1940s; the exigencies of the Berlin Airlift 1948-1949 extended it and the only fuel available was low-octane co-operative branded Pool, introduced as a wartime emergency. The motor trade was in disarray. Cars of the 1930s were worn out; hardly any had been made for five years so as production restarted the demand was insatiable. The war had to be paid for, and the incoming radically-disposed Labour government imposed restrictions against guarantees to export. Steel was assigned only to car makers who exported 8 out of every 10 they made. It was a traditionally class-conscious administration that never really got on with cars.

Sir Stafford Cripps, President of the Board of Trade and Chancellor of the Exchequer from 1947-1950, nearly did for the entire British luxury car market. His counterparts in France had already annihilated its, bleeding Delahaye, Delage, Gregoire, Hotchkiss and Talbot to death through taxation and regulation. Cripps levied Purchase Tax and Double Purchase Tax as well as imposing niggardly steel allocations. MG, along with Daimler, Armstrong Siddeley, Jaguar and Rover only survived because it was earning dollars, so the austere Cripps had to put up with it. His egalitarianism saw no place for luxury goods of any sort, let alone showy sports cars and so far as the Nuffield Organisation was concerned, MG could be consigned to a backwater.

Undaunted however, within weeks of the war's end, Abingdon set up a car production line. Each bare chassis was pushed along by hand, acquiring components and bodywork as it went. On the face of it MGs seemed unlikely to make headway abroad yet among returning United States servicemen especially, demand for small sporty British cars persisted. Production started tentatively; like most of the industry MG had little choice but take up where it had left off with whatever it had been making in 1939. Between 1946 and 1949 it sold 1820 TCs to America and almost as many in Australia. New cars were nearly unobtainable on the domestic market save for priority buyers such as doctors and workers in what were termed essential businesses. Second-hand prices went virtually out of control so a covenant scheme had to be introduced to prevent profiteering. In the middle of 1946 TC prices went up from £375 to £412, inflated further by Purchase Tax to £528.

In a sign of the Nuffield group's low expectations for MG, production of Rileys was moved to Abingdon. Acquisitive as ever, Lord Nuffield had personally bought Riley, "As old as the Industry…" in 1938 from receivership for £143,000, transferring ownership to Morris Motors Ltd for £100. Progressively engineered with a notable twin-high-camshaft engine and successful in motor racing, often against MG, Riley's reorganisation had barely begun before war broke out. Victor Riley remained nominally in charge until 1947 when he was dismissed and its Durbar Avenue Coventry factory was constrained only to making engines. A move to shift MG there was narrowly averted by Propert and John William Yates Thornley (1909-1994) who had taken over as sales and service manager in 1947.

MG investment still languished. Nuffield-Morris was predicated on the largest volume production possible. Keeping down unit costs was vital to its prosperity and its economies of scale seemed to preclude a sports car maker as unlikely to achieve them. In despair the big organisation decided to make the best of MG and use small-volume production for premium-priced projects. If it could not achieve big numbers it would try Kimber's thesis that it was always possible to charge 20 per cent more provided a car was 10 per cent faster, looked 10 per cent better and only cost about the same to make. It could be profitable. It was not an easy proposition to put before the parsimonious Lord Nuffield but Kimber had made it work in the 1920s and it could still work now. Nuffield management knew it applied to top-end luxury saloons although extending it to harshly sprung sports cars seemed unlikely. It saw the prosperous Belgian market importing soft-riding 6-seat Buicks with Dynaflow automatics, air conditioning and radios and concluded that MGs, even new TCs with 4in (10.2cm) more elbow-room, belonged to history.

Instead it wanted to turn Abingdon into a satellite factory, making the stylish torsion bar sprung 1½Litre Riley RMA. So from 1946 alongside TCs it turned out 5479 of the vogueish modern saloons (below), replacing it with a tidied-up RME in 1952 of which 3446 were made. The RMB 2½Litre was also made at Abingdon on a similar chassis, and between 1946 and 1953 out of a total 6900, MG made

5639 of them. In December 1950 a Riley was the 10,000th car produced at Abingdon, which also made 386 of 507 RMC Roadsters and 498 of the 502 RMD drophead coupes.

When Gerald Marley Palmer (1911-1999) designed the Riley Pathfinder of 1953-1957 on a chassis shared with the Wolseley 6/90, the little factory made all 5152. A further reorganisation of Nuffield led to Wolseley closing its traditional home at Ward End, Birmingham and transferring production to Cowley. When the Riley 1.5 was introduced with Morris Minor suspension and steering in 1957, Abingdon made the first 150. Once sure of its success, production was transferred to Cowley along with the lugubrious Riley 2.6, which abandoned its classic 4-cylinder Riley engine and settled for a Wolseley 6-cyinder that matched its lifeless cam steering. MG made 434 of the 2000 produced before the model passed into oblivion in 1959.

The talented Palmer had grown up in Bulawayo, son of Rhodesia's chief railway engineer, and taken an apprenticeship at Scammel in the 1930s. Aged 27 he showed Cecil Kimber a sports car he created. Unable to offer him a design position at Abingdon, which did not have its own creative department, Kimber liked it sufficiently to suggest Palmer develop his talent as a design draughtsman at Morris Motors. Like an architect who extended his strategies to furnishings as well as structures, Palmer instead left for Jowett and masterminded one of the most radical cars of the post-war period, the 1.5litre Javelin. He planned the entire car with a stylish body and developed the Bradford company's primitive flat-twin engine into a robust 4-cylinder. The Javelin had torsion bar suspension, precise rack and pinion steering, handled like a thoroughbred and won its class in the Monte Carlo Rally.

It was not, alas, enough to be Britain's most sophisticated family car ahead of anything from Continental Europe. Jowett had over-reached itself, only 23,307 Javelins were ever built and the company failed after Ford took over Briggs Motor Bodies. Palmer was coaxed back to an uncomfortable career at BMC and even though his MG Magnette and Wolseley 4/44 were well-received and successful he found his work difficult through company in-fighting. His talent and energy were largely ignored by belligerent bosses, he fell out with Alec Issigonis and left to join Vauxhall.

In 1953 John Thornley was promoted MG's general manager in succession to Jack Tatlow. He had been at MG since the 1930s, was as dedicated to the Kimber legacy as LP Lord had been to erasing it and was first and foremost a sports car enthusiast. Lord would have liked to expunge Nuffield too, out of animosity following dismissal after demanding a share of the business from Morris; however he was pragmatic enough to keep MG in business even though he had no time for Barré Lyndon-style romanticism. MG was always better run by strong-minded self-confident individuals. Administration by committees or parsimonious accountants invariably proved disappointing. Success came through practicality, style and good value. Kimber's skill, which he wove into MG culture, lay in knowing his market and securing the means to meet it.

Thornley backed himself up with a formidable array of engineers, notably Albert Sydney Enever (1906-1993), whose role in MG's survival never quite gained the recognition it deserved. Between joining in 1930 until retiring in 1971 as chief engineer, Syd Enever's influence on MG cars was profound. His qualifications were more practical than formal, and he was accorded less than his due by management's regarding him more as a skilled artisan than the brilliantly intuitive engineer he was. EX135, the Gardner MG, and two of the most successful sports cars ever, the MGA and MGB, were Enever inspirations.

MG had evolved not as a stand-alone project, but always as an adjunct of a bigger player. Kimber had made the best of dependency on Morris/Nuffield but following the unlikely merger of Austin with the Nuffield Organisation in 1952 to form the British Motor Corporation (BMC), the situation changed.

The effects of BMC's creation were profound. Austin and Nuffield had shared two-fifths of motor sales in Britain. They had been competing in cars and commercial vehicles for more than four decades. There was intense rivalry not only between the corporate bodies but also between their engineers, individuals and salespeople. Cultural and commercial differences were profound, synergies indistinguishable; Austin was dependable, robust, middle-class and self-contained at Longbridge, Birmingham while Nuffield sprawled through acquisitions by William Morris, disparate and materialist across half the Midlands. On the face of it the BMC could have been one of the most formidable car manufacturing companies in the world. In hindsight it was the creation of a management elite enriching itself, setting back the United Kingdom's motor industry by a generation and bound it for disintegration.

In April 1952 BMC took up the shares of Austin and Morris Motors, the holding company for Nuffield-owned MG, Riley, Wolseley and Morris. It was one thing creating a corporate structure for the new corporation; unfortunately Austin people and Morris people never got on. Neither company had been well managed. Lord Nuffield became the first chairman but by the end of the year handed over to Sir Leonard Lord who remained until he turned 65 in 1961, two years before the death of Viscount Nuffield. Lord shared responsibilities with George Harriman of Austin from 1956 and BMC's headquarters were set up within those of Austin in the extensive complex of Longbridge, where they had been making cars for more than half a century.

Practicalities following the merger led to abandonment, after only three years, of all Morris's rather dowdy old engines, produced in scattered run-down factories. Design work was left to the Austin Drawing Office under projects coded ADO which, it was now said, stood for Amalgamated Drawing Office. Longbridge was well integrated, by contrast with Nuffield's 16 often dated plants dispersed throughout the English Midlands, shuttling lorry-loads of components between them. Nuffield's sales and dealer network and cost controls were probably the better, but from the outset BMC's market share went into decline. Quality control was no match for imports from the Far East over the following decade with cars that came with built-in radios, worthy warranties and bodywork that didn't rust.

The outlook for MG grew bleaker following Lord's endorsement of the Austin-Healey 100 at the 1952 London Motor Show, with plans to make it at Longbridge. The move coincided with the refusal by the new BMC to back a replacement for the TD. Lord was remembered for his abrupt closure of

Abingdon's competition programme in the 1930s and the Austin-Healey was regarded as opportunist, using up 2.6litre engines and transmissions left over from the failure of Austin's A90 Atlantic. That had been an initiative of Lord's to gain market in America where MG had been moderately successful and although well-intentioned it was clumsy. Now, however, with the advent of the stylish and faster Healey, sales of MG TDs halved. Against the perfectly proportioned sleek 2-seater the upright old-fashioned MG was doomed. To make things worse Y-series collapsed too. Instead, strapped for cash, MG cobbled together the TF greeted, curiously as it seemed later, with some disdain. Like Lord's ungainly A90 it would be generations before TFs came to be regarded as covetable classics.

It was consequently a surprise, although welcome, to find a measure of reinstatement for Abingdon coming unexpectedly from the Austin side of BMC. Harriman and Lord decided to encourage BMC into motor sport. The MGA was reluctantly sanctioned to prototype stage and committed to Le Mans for 1955. Confidence in the car market was growing. Post war difficulties, steel shortages, rationing and government-imposed austerity were drifting into the past. A meeting in December 1954 set up a new BMC Competitions Department; the old Nuffield Competitions Committee was disbanded. Optimism returned.

In point of fact the corporation had been obliged to raise its game. Triumph in Coventry was planning more exposure for the TR2 that had just won the RAC Rally. It was setting up a competitions department under Ken Richardson, formerly with BRM. Elsewhere Jaguar was setting the pace with the XK120 and winning Le Mans outright with the C-type. Rootes had been putting up solid performances in rallies, especially the Alpine, using Mike Hawthorn, Peter Collins and Stirling Moss who won three Alpine Cups in succession and became one of only five ever to win a Gold Cup. Rootes had a lighter, prettier and swifter Alpine planned as well as a Sunbeam version of the Hillman Minx with disc brakes and overdrive. This was the stylish Rapier and a team of them under the direction of Norman Garrad was shaping up well. BMC did not want to be left out.

John Thornley came under pressure to create a competitions programme based at Abingdon. MG had form. It knew about motor sport and if it could make Rileys it could in due course make Austin-Healeys. Thornley consulted Sydney Charles Houghton "Sammy" Davis (1887-1981), sports editor of *The Autocar*, former Le Mans driver and doyen of motor racing. Thornley was searching for somebody to run it and Davis recommended Marcus Mordaunt Bertram Chambers (1910-2009).

It was an inspired appointment. Son of a naval captain who spent his early life abroad, Chambers survived the munitions explosion that wrecked Halifax, Nova Scotia in December 1917. His family returned to Britain, by 1938 he was running his own garage in Chichester and driving Bentleys in

Vintage Sports Car Club events. He had come 10th in a works HRG at Le Mans in 1937 and 13th, winning his class, in 1938. After war service in the RNVR Chambers resumed his connections with HRG but in 1948 realised the limitations of amateur drivers paying their own way in international sports car races.

BMC's new Competitions Department was given a generous budget of £100,000 (the equivalent of £2.5 million in 21st century money), which meant that resources would not be a problem. In 1955 it was augmented with £10,000 from Castrol. Chambers set up a small office next to the development department at Abingdon and entered the Monte Carlo Rally with MG Magnettes and Austin Westminsters. "The tail-happy MGs required first aid at Caffyn's dealership in Ashford on the run to Dover, when the string binding of the rear springs was cut to allow more movement," Chambers wrote later in a book *BMC Competitions Department Secrets* chronicling its remarkable successes. Not all the corporation's cars were suited to competition but it was part of his job to influence new model plans and he suggested a GT MG Magnette for rallying with a twin-cam engine, alloy panels, knock-off wheels and a big fuel tank giving a 300-mile range. The nearest he got was one made for Nancy Mitchell in the 1956 Tulip Rally until a scrutineer excluded her after running a magnet over it and discovering much was made of aluminium.

In 1955 Stirling Moss's sister Pat, a successful horsewoman, wanted to drive a Triumph TR2 but Chambers persuaded her to compete with an MG TF instead. She took part in the 1955 RAC Rally, BMC publicity liking the idea so much it printed Ecurie Safety Fast rosettes (left) like the ones Pat was accustomed to winning at showjumping. Chambers also recruited Ian Appleyard, Britain's most successful rally driver at the wheel of Jaguar XK120s. "With the emphasis switching from punch between corners to agility around cones and nimble handling, Ian now reckoned a smaller car with less power could do well … and fancied an MG YB. I talked him out of that and found another TF, in which Mr and Mrs Appleyard finished eighth overall. Triumph's Competitions Department cleaned up with the Standard Ten, finishing first and third, using some BMC bits."

Nancy Mitchell won the Coupe des Dames in the Alpine Rally in an MGA, while Pat Moss and Ann Wisdom won the Liège-Rome-Liège in 1960 in the hot and noisy but increasingly competitive Austin Healey 3000. It was the first outright win for a women's crew in a major high-speed European rally. Donald and Erle Morley were also successful with the big Healey and were among the first to ever develop pace notes. BMC drivers earned around £25 per event, plus expenses.

By the time Chambers took charge MG was already committed to Le Mans for 1955, which turned out to be a turning point affecting the whole of motor sport. The accident in which Pierre Levegh's Mercedes-Benz 300SLR was launched over Lance Macklin's Austin-Healey killed over 80 and shocked the world. Switzerland banned motor sport, including its own long-established grand prix at Berne and questions were raised about the validity of specialist prototype racing cars made of exotic materials such as magnesium alloys, blamed for some of the explosive results of the crash. Le Mans was reorganised as a display event for production based sports cars, which BMC urged the RAC to encourage. In the event the pits complex and track layout were changed, some of the rules were indeed altered but in a few years the status quo was effectively restored.

Le Mans 1955 was expected to be a showdown between Mercedes-Benz and Jaguar but so far as MG was concerned it came down to rivalry with Triumph. Marcus Chambers: "Triumphs and MGs had a race of attrition. After the big accident smoke rose over the grandstand. Dick Jacobs and MG number 42 went missing and another column of smoke rose from White House Corner. Dick was seriously injured. The previous accident had caused a momentary lapse of concentration, and the MG had hit a bank. After three days in hospital – his life saved by an MG timekeeper who happened to be a doctor – Jacobs was flown home on the personal instruction of BMC Chairman Sir Leonard Lord, but was not discharged from hospital until October. As for the race, the new MGs had done well, despite the accidents overshadowing everything. The surviving MGs finished 12th and 17th; the best Triumph was 14th, Ken Richardson's TR 15th, and 19th place was taken by a TR2 driven by Morris-Goodall."

One result of the accident was to divert Chambers' attention and BMC competitions policy away from MG racing. "In the following Ulster TT, two of the Le Mans MGs were entered with experimental twin-cam engines: one designed at great cost by Austin, and the other designed at great cost by Morris Engines; now the rivalry of the two wings of BMC would be made public because the engines were to race against each other. The idea behind this was that the better engine would go into production as a special version of the MGA. During the TT two drivers were killed, several cars were involved in accidents and press was again filled with gory pictures and headlines containing the word 'tragedy'.

As a result Sir Leonard Lord decreed that motorsport efforts were to be concentrated on rallies and record-breaking."

Production practicalities at Abingdon were gradually turning round. In 1956 a design office was re-established with Don Hayter as senior layout draughtsman and the 100,000th MG, a left-hand drive MGA 1500 tourer was built. The factory's role within BMC was bolstered in January 1957 when Donald Mitchell Healey (1898-1988) demonstrated his prototype Healey 100 to George Harriman. A former Royal Flying Corps pilot, rally driver and leading engineer at Triumph in the 1930s, Healey had a lifetime in the motor industry, making cars under his own name, notably with Riley engines. What became the Austin-Healey BN1, of which 10,688 were made under licence by BMC from 1953 at Longbridge, went on to critical and sales success, and 50 of a specialist 100S assembled at The Cape, Warwick from 1954. The BN2 derivative was introduced in 1955 of which a number were developed to be 100M models. The body and chassis of these was made by Jensen at West Bromwich.

In the autumn of 1956 the wheelbase was extended to a 2+2 version with a C-series 6-cylinder instead of the old A90 engine to make the 100-Six BN4. The following year, while activity at Warwick was wound down, Donald and his son Geoffrey became consultants and production was moved to Abingdon. Other derivatives followed, including three Marks of the Austin-Healey 3000, until 1967. The conception was still pure Kimber, the entrepreneurial and engineering skills of the Healey family building prototypes on their own initiative, seizing opportunities to exploit them like a freelance development department and selling manufacturing rights almost before BMC realised they were needed.

Donald Healey was also working on a small low-cost sports car following conversations with Sir Leonard Lord in 1956. A scheme for one was drawn up in 1952 by his body designer, Gerry Coker, using an Austin A30 engine. Lord was open to a meeker alternative to the Austin-Healey 100 after North American dealers told him that its power and speed put some customers off. Three big Healeys out of four were being written off in accidents.

Above: Austin-Healey 3000 Mk III of 1966 towards the end of a successful series of sports cars

The result was the Austin-Healey Sprite Mark I and Austin co-operated wholeheartedly with the Healey design consultancy. It ensured components were available for prototype development and there was even a liaison official to supply drawings and specifications of the A35 front suspension, engine and axle. All the Healeys had to do was fill in the bits between. Some proprietary units were unsuitable, like the A35's mechanically operated rear brakes. Likewise its Cam Gears steering box, with ball-joints and idler arm was not precise enough for a sports car, so a Morris Minor rack-and-pinion was used instead. The clutch was operated hydraulically through an MG twin master-cylinder and an MG manifold for the twin SU carburettors replaced the Austin's single Zenith. The Healeys, Donald and son Geoffrey, were following a layout exemplified by the D-Type Jaguar, with a stiff centre structure and outriggers at the front carrying the engine and front suspension. The rear axle trailed behind so loading front and back were fed into the centre portion rather than the tail, which could be built as a lightweight shell.

A straight platform floor ended just behind the seats and a deep, enclosed propeller-shaft tunnel joined the scuttle pressings at the front and a sloping transverse panel at the rear ahead of the axle stiffened it. Door sills were 5in (12.7cm) deep, fully boxed, and longitudinal-section pressings were welded to the floor underneath the seats. At the rear of the platform these were the take-up points for the suspension and in front they met a beam running the width of the car under the driver's knees. This formed the main cross spar on which the scuttle structure and the front longitudinals were mounted.

Body panels were simple. Most were flat; th only double-curvatures were in the tail pressings forming the hollow shell, light through being relatively unstressed and gaining strength through not having a boot aperture. Luggage space was reached through a yawning gap behind the seats.

Left: The author's Mark 1 Austin-Healey Sprite at Turnberry, Ayrshire.

DB2 Aston Martins and Lotus 11s both had entire fronts that opened to provide access to the engine and front suspension but the entire bonnet structure of the Sprite, complete with grille and headlights, lifted on to supporting struts. The body was a masterpiece of simplicity and practicality even though Coker's 1952 plan for pop-up headlights operated by rods and levers was abandoned. Panel Craft of Birmingham made two prototypes, which were essentially carried through to production cars.

Floor pan and scuttle, front suspension and the location of major components survived for some 21 years. In Mark I form, with bulbous headlights in the bonnet top giving it a pop-eyed frog-like look, the Sprite cost little more than contemporary small saloons. It was economical, it gave its driver a lively ride on its stiff quarter-elliptic rear springs, but it had all the feel and precision of a thoroughbred and was produced at Abingdon and sold through Austin dealers. Final development work included tortuous fatigue tests on pavé that revealed weaknesses that were put right from its launch in May 1958 right through to May 1961 when the Mark II Sprite was introduced along with an MG Midget counterpart.

Don Hayter, charged with drawing up Abingdon's next big project had a long motor industry pedigree. He started in 1942 at Pressed Steel on aircraft then went to Aston Martin from 1954-1956 arriving at MG as chief body draughtsman, basing the design of the MGA Twin Cam Le Mans Coupe that ran in 1960 on his experience with Aston's DB3S. By June 19 1958 Hayter and chief body engineer Jim O'Neill were engaged in the MGA's replacement taking the next EX number 214 to define the shape of the MGB.

Frustrated by indecision over what engine it would have, the overall design was approved by Syd Enever in 1960. Anything less than 2.0 litres would feel rather meek in a market against Triumph TR2s or Austin-Healey 100s. The BMC B-series in the MGA was no match. It had been designed in 1943, developed in various forms for Austins (including the ill-starred A90) there was little prospect that the Austin end of BMC would allow it to be in anything as fast as an Austin-Healey 100. Revised in 1953 to 1,489cc for the Magnette by enlarging its bore from 65 to 73mm it was all the management had to offer. Narrow angle (60deg) V4s and V6s under development at Jaguar, not yet within BMC, were considered but abandoned owing to vibration problems although the front bonnet access had been designed with them in mind.

Syd Enever secured options on a compact V8 designed by Edward Turner of Triumph motorcycle fame, inherited by Jaguar with the Daimler SP250. A 90deg V8 2½litre it was used successfully in a

Jaguar/Daimler saloon 1963-1969 and Enever contemplated it even after production had started with the 68bhp (50.7kW) B-series that made the B only a little faster than an MGA. Alex Leslie Hounslow (1911-1976), MG's long-serving development engineer and famously Nuvolari's riding mechanic of 1933, successfully put a Turner V8 in an early production car without, said Hayter, much alteration to the MGB's structural side-members or suspension supports. The upshot was that in years to come bigger engines were easily accommodated in a stout monocoque always intended to leave room for more power.

The punishing international rally schedule put Marcus Chambers on the road with the BMC team for six months of the year. By 1961 he was tired of the long days, left for a job as a service manager but returned to motor sport after two years to make the Sunbeam Tiger competitive for the Rootes group.

John Thornley was tasked with finding a replacement for Chambers. Stuart Turner: "After a brief interview I was invited to become BMC Competitions Manager. I didn't know I was even in the frame for the job. Looking back, maybe a mild reputation for 'rallymanship' helped; not cheating, but pushing the regulations to the limits, although I'm not sure John Thornley was convinced I was the man for the job and think it was really an endorsement by Marcus that clinched it. As an avid enthusiast, going to work at the MG factory at Abingdon, where the famous octagon featured everywhere, was like a dream. I was 28, and suddenly found myself working with industry legends, including Alec Hounslow. And they were going to pay me £1250 a year to do so. Magic!"

Turner was probably the best of the three managers BMC "Comps" ever had. Unencumbered by Chambers' inhibitions about employing only British drivers for the British Motor Corporation after sitting beside Erik Carlsson (Saab) when he won the 1960 RAC Rally Turner was convinced that Scandinavians' experience of driving on snow and ice equipped them uniquely for the special stage events becoming customary. His recruitment of Rauno Aaltonen and Timo Makinen alongside Paddy Hopkirk provided BMC with countless victories in major events convincing not only BMC management but also the wider public of the Mini's astonishing performance.

Following its great triumph in the 1964 Monte Carlo Rally Turner wrote: "The Monte result was a real team effort … When I got back to Abingdon I found the flags flying and ribbed John Thornley that at last MG people had recognised the Mini. He told me the flags were flying because the Morley brothers had won the GT category in an MGB."

Interest in the MG side of "comps" was not allowed to atrophy despite its concentration on rallying. The MGB was developed and despite its essential roadworthiness it could still turn in good performances. Paddy Hopkirk wrote in *Autosport* following one of them.

"Maybe if Le Mans had been 48 hours, we might have won it. I've never seen a race with so many (or so few) sick and tired motor cars running at the end. Our MGB owed a lot to its Abingdon preparation. You could have eaten your breakfast off the engine and even at the end there were no oil leaks or anything like that. The car was much the same as last year with a few extra aluminium panels at the rear. It would do 126mph (202.8kph) on Mulsanne without a tow, but towards the end when the Sears/ Thompson Cobra was slowed with low oil pressure we could slipstream with it up to about

132mph (212.4kph), which took around 4sec a lap off our usual 4min 50sec. Our timekeepers could hardly believe their stopwatches in the last couple of hours when we suddenly started turning in times quicker than we had at the start. Apparently our car was so unracerlike that the crowds looked forward to it every time it came through the Esses. It heeled over and squealed like a touring car. Inside, it was just like any other MGB with carpets, trim, door handles and everything. The only thing it hadn't got was a radio.

"Unscheduled pit stops held us up. The R7s with which we began the race rubbed on the body and Andrew Hedges, who shared the car with me, had a blow-out and drove threequarters of a lap on the flat. For some reason, this damaged the exhaust and the stewards called us in to wire it up. We changed to R6s, which are narrower, and actually faster on the MGB because they have a lower rolling resistance.

"One thing about driving a slow car at Le Mans is you certainly get time to watch the race. You spend a lot of time looking in the mirror, but at least one Ferrari driver appeared to think you should pull into a lay-by every time he wants to pass. The only other trouble during the race was four laps from the end, when we were trying to catch up Graham Hill and Jackie Stewart in the Rover-BRM. We were making up 21sec a lap as they got slower and we had 10th place in our sights when I was baulked on a corner by a Ferrari, braked heavily, and burnt off what was left of our front pads. This broke a piston seal in one brake caliper leaving me with only the handbrake."

It was a remarkable performance for a relatively heavy and nearly standard car. V12 3.3litre Ferrari 250LMs were 1st and 2nd a 275GTB 3rd; Porsches were 4th and 5th; another 250LM 6th; 365 4.4litre 7th; Shelby Cobra 8th; Iso Grifo 9th, the Rover-BRM turbine car 10th and the MG 11th. Paul Hawkins and John Rhodes came 12th, another Sprite driven by Rauno Aaltonen and Clive Baker was 16th while the Triumph Spitfires managed 13th and 14th.

The 1950 10,000th car at Abingdon, a Riley, the 1956 design office established under Don Hayter, and the 100,000th MG, an MGA 1500 were factory milestones. Austin-Healey production flourished and the first handful of Riley 1.5s in 1957 secured diversity well into the 1960s. More security came with the Magnette of 1953 and, although a close relative of the Wolseley 4/44 which was made at Cowley, Magnette ZA and ZB production was secured for Abingdon. The pattern stretched Abingdon. By the dawn of the 1960s as its sports car production was about to be improved by the exemplary MGB and he reputation of the Magnette was at its zenith BMC collectivism decreed that its Mk III would be a clone of the infamous Farina-designed badge-engineered cars that signalled the start of a 20-year decline in the fortunes of the indigenous Midlands motor industry.

Classic Cars

20th Anniversary 1973 - 1993 MG TC, as featured in Classic Cars, June 1974

MG Classics Model by Model 2

1945-1965

1945-1949 TC Midget

As soon as war work ended, with Kimber gone and commercial and industrial recovery still a long way off, there was pressure to get production under way. There was certainly no time to develop a new model and with a diminished Nuffield range from which to draw components the luxury SA, VA, and WA were abandoned. The T-series was simplified as a matter of expediency so the sliding trunnions, a cherished feature of Midgets from the Montlhéry C-type of 1931, were sacrificed. The body was made 4in (10.2cm) wider and in October 1945 production started, but only 81 were made before the end of the year. The following year there were 1675. It was a seller's market both in Britain and overseas and MG was being exhorted, indeed compelled by government diktat, to export as many as possible. Notwithstanding that, the car's success was astonishing despite its disdain of what were regarded as essential attributes in markets such as the United States. The TC had no heater, it was obtainable only in right hand drive and the ride was distinctly firm. There was only 4in (10.2cm) of seat cushion adjustment, less for the backrests, although the steering column had a telescopic arrangement to alter the reach. Yet by 1949 MG had made and sold 10,000, which was more than three times the total of any MG model hitherto.

Body sports, 2 doors, 2 seats; weight 1735lb (787kg). Engine 4 cylinders, in-line; front; 66.5mm x 90mm, 1250cc; cr 7.25:1; 54.4bhp (40.6kW) @ 5200rpm; 43.5bhp (32.4kW)/l; 63.5lbft (85Nm) @ 2600rpm. Engine Structure duplex roller chain-driven camshaft; pushrod overhead valve; cast iron block and top half of crankcase; aluminium ribbed sump; cast iron head; coil ignition; two semi-downdraught SU carburettors; 3-bearing crank; cooling by pump, fan, and thermostat. Transmission rear wheel drive; Borg & Beck single dry plate clutch; 4-speed manual gearbox; synchro on second, third, and top; spiral bevel final drive 5.125:1. Chassis Details parallel steel channel-section frame, underslung at rear, with four tubular cross-bracings, and boxed round engine and gearbox; semi-elliptic springs front and rear with shackles; Luvax lever arm hydraulic dampers; Lockheed hydraulic brakes with 9in (22.9cm) drums; Bishop cam gear steering; 13.5gal (61.4l) fuel tank; 4.50 x 19.0 tyres; centre-lock wire wheels. Dimensions wheelbase 94in (238.8cm); track 45in (114.3cm); turning circle 37ft (11.3m); ground clearance 6in (15.2cm); length 139.5in (354.3cm); width 56in (142.2cm); height 53in (134.6cm). Performance maximum speed 75mph (120.7kph); 15.9mph (25.6kph) @ 1000rpm; 0-60mph (96kph) 22.7sec; 14.5kg/bhp (19.4kg/kW); fuel consumption 28mpg (10.1l/100km). PRICE sports 2-seater £527.16s 8d (£527.83) including purchase tax. production last car 10 November 1949 serial number 10,001, chassis 10,251.

1947-1951 YA 1¼ litre saloon

Independent front suspension, rack and pinion steering instead of Bishop Cam, XPAG engine and part-monocoque body construction; the ingredients of post-war MGs were coming together. An MG Ten planned for 1940 had been cancelled and the Y-type fulfilled a requirement during severe petrol rationing for a small economical luxury saloon, not swift but handling and riding well. The coil spring and wishbone ifs was drawn up by Alec Issigonis as a young designer and Jack Daniels for a Morris Ten in the 1930s but the management chose something cheaper. Instead it was incorporated in the YA and essentially survived for the best part of half a century. The Y-type was based on Morris Series E and Wolseley 8 press-work at Nuffield Metal Products in Birmingham, although so subtly done the resemblance almost went unnoticed.

MG had acquired a nostalgic quality in wartime and Y-types backed up the sports cars in club navigational rallies that were supplanting trials as the leading variety of popular motor sport. Geoff Holt, driving a TD, won the first Daily Express 1000 mile rally in 1950, and Len Shaw came third in a YA, both winning their class. Production of the YA began in April 1947 until superseded by the YB.

A few chassis were sold to Swiss coachbuilders, and a series of special-bodied cars was planned by Roger Barlow of Los Angeles, one of MG's flourishing export markets.

BODY saloon; 4 doors, 4 seats; weight 21cwt (1067kg). ENGINE 4 cylinders, in-line; front; 66.5mm x 90mm, 1250cc; compr 7.25:1; 46bhp (34.3kW) @ 4800rpm; 36.8bhp (27.4kW) /l; 59lbft (79Nm) @ 2400rpm. ENGINE STRUCTURE duplex roller chain-driven camshaft; pushrod ohv; cast iron block and top half of crankcase; aluminium ribbed sump; cast iron head; coil ignition; SU 1¼in carburettor; 3-bearing crank; cooling by pump, fan, and thermostat. TRANSMISSION rear wheel drive; 7¼in Borg & Beck sdp; 4-speed manual gearbox; synchromesh on second, third, and top; Hardy Spicer needle roller bearing prop shaft; spiral bevel final drive 5.142:1. CHASSIS DETAILS steel welded box-section side rails and tubular cross members, underslung at rear; ifs by coil spring and wishbone, live rear axle with shackled semi-elliptic leaf springs and Panhard rod; Luvax-Girling lever arm hydraulic dampers; Lockheed hydraulic brakes with 9in (22.9cm) (22.7cm) drums; rack and pinion steering; 13.5gal (62l) fuel tank; 5.25 x 16 tyres; bolt-on steel disc wheels, Jackall hydraulic jacking. DIMENSIONS wheelbase 99in (22.9cm) (251.5cm); track 47.4in (120.4cm) front, 50in (127cm) rear; length 161in (409cm); width 59in (22.9cm) (149.8cm); height 58in (20.3cm) (147cm). PERFORMANCE maximum speed 71mph (114.3kph); 0-60mph (96kph) 28.2sec; 23.2kg/bhp (31.1kg/kW); fuel con 27mpg (10.5l/100km). PRICE £671.11s 8d (£671 58) including purchase tax. PRODUCTION 6158.

Right: Y-type saloon had spare wheel under the boot behind detachable panel

1948-1950 YT 1¼ litre open 4-seater

Introduced at the first post-war motor show in Earls Court, London the YT looked more 1930s than 1940s. Vaguely more sporting than the saloon, it had small cowls on the scuttle and a TC-style fabric-covered instrument panel instead of its counterpart's veneered wood. There were some detail differences from the underpinnings of the saloon on account of the tourer's lighter weight. The rear springs were dissimilar and there were alterations to the steering rack and column. Both doors were deeply cut away in support of its sporting nature but this meant detachable sidescreens instead of the winding windows of its pre-war forebears. Hood and frame were stored in a compartment that reduced space in the back seat. Under pressure to export, the tourer was available with left hand drive and even had flashing direction indicators instead of the saloon's semaphore trafficators. Since only a few were expected to be sold in Britain (the bulk of production went to America and Australia) it was priced the same as the saloon on the home market. The YT did not benefit from modifications carried out on the Y-series at the end of 1951 coinciding with the introduction of the TD. Even with the TC's twin carburettor engine the YT was not a success, an indication that the world had moved on from the leisured, unhurried era of the VA tourer.

BODY open tourer, 2 doors, 4 seats; weight 2184lb (991kg). ENGINE 4 cylinders, in-line; front; 66.5mm x 90mm, 1250cc; compr 7.25:1; 54bhp (40.3kW) @ 5200rpm; 43.2bhp (32.2kW)/l; 63.8lbft (92Nm) @ 2600rpm. ENGINE STRUCTURE duplex roller chain-driven camshaft; pushrod overhead valve; cast iron block and top half of crankcase; aluminium ribbed sump; cast iron head; coil ignition; two SU 1¼in H2 carburettors; 3-bearing crankshaft; cooling by pump, fan, and thermostat. TRANSMISSION rear wheel drive; 7¼in Borg & Beck single dry plate clutch; 4-speed manual gearbox; synchromesh on second, third, and top; Hardy Spicer needle roller bearing prop shaft; spiral bevel final drive 5.142:1. CHASSIS DETAILS steel welded box-section side rails and tubular cross members, underslung at rear; independent front suspension by coil spring and wishbone, live rear axle with shackled semi-elliptic leaf springs and Panhard rod; Luvax-Girling lever arm hydraulic dampers; Lockheed hydraulic brakes with 9in (22.9cm) (22.7cm) drums; rack and pinion steering; 13.5gal (61.4l) fuel tank; 5.25 x 16 tyres; bolt-on steel disc wheels, Jackall hydraulic jacking. DIMENSIONS wheelbase 99in (22.9cm) (251.5cm); track front 47.4in (120.4cm), rear 50in (127cm); length 161in (408.9cm); width 59in (22.9cm) (149.8cm); height 58in (20.3cm) (147cm). PERFORMANCE maximum speed 76mph (122.3kph); 0-60mph (96kph) 28.2sec;18.3kg/bhp (24.6kg/kW); fuel consumption 27mpg (10.5l/100km). PRICE £671.11s 8d (£671 58) including purchase tax. PRODUCTION 877.

1947 Coachbuilt Y-types

Above: Zagato-bodied YA

Back in the 1920s MGs had been essentially special bodies on a separate chassis. Cecil Kimber's own car was a supercharged K1 with bodywork by a specialist in rakish coupes, Corsica, which took its name from the Kings Cross street where they were made, not anything Mediterranean. Magna bodies were made by Swiss coachbuilders Sportcar AG, and a P-type by Hami also of Zürich. Swiss Keller had made a WA look like a Mercedes-Benz 1930s parade car, while Salmons' one was more like a Bentley. By the 1940s many car manufacturers had turned to unitary monocoque structures, making the Y-type a rare old-style chassis, so it was seized upon by grateful Continental coachbuilders. Pininfarina proved, with its ground-breaking 1948 Cisitalia, that it was possible to set a styling trend for a generation even if the make of car on which it was set had been in decline. Zagato was ambitious and showed a YA at the Lugano Salon in 1949, which was also something of a mould-breaker but scarcely a trend-setter. Its cockpit was composed of curved-glass panels in the style of a wartime fighter aircraft canopy. It must have given an airy ambience to the interior, unlike the rather gloomily trimmed YA saloon. Anyway the idea did not catch on and only one was ever made. Castagnata, another Italian carrozzeria from Milan, made a more elegant attempt with a theme taken up later by Nuccio Bertone for the more commercially successful American Arnolt-MG.

BODY coupe, 2 doors, 4 seats; weight approx 20cwt (1016kg). ENGINE 4 cylinders, in-line; front; 66.5mm x 90mm, 1250cc; compr 7.25:1; 46bhp (34.3kW) @ 4800rpm; 36.8bhp (27.4kW)/l; 59lbft (79Nm) @ 2400rpm. ENGINE STRUCTURE duplex roller chain-driven camshaft; pushrod ohv; cast iron block and top half of crankcase; aluminium ribbed sump; cast iron head; coil ignition; SU 1¼in carb; 3-bearing crankshaft; cooling by pump, fan, and thermostat. TRANSMISSION rear wheel drive; 7¼in Borg & Beck single dry plate clutch; 4-speed manual gearbox; synchromesh on second, third, and top; Hardy Spicer needle roller bearing prop shaft; spiral bevel final drive 5.142:1.CHASSIS DETAILS steel welded box-section side rails and tubular cross members, underslung rear; independent front suspension by coil spring and wishbone, live rear axle with shackled semi-elliptic leaf springs and Panhard rod; Luvax-Girling lever arm hydraulic dampers; Lockheed hydraulic brakes with 9in (22.9cm) (22.7cm) drums; rack and pinion steering; 13.5gal (61.3l) tank; 5.25 x 16 tyres; bolt-on steel wheels, Jackall hydraulic jacking. DIMENSIONS wheelbase 99in (22.9cm) (251.5cm); track front 47.4in (120.4cm), rear 50in (127cm). PERFORMANCE maximum speed 71mph (114kph); 0-60mph (96kph) @ 28.2sec; 22.1kg/bhp (29.6kg/kW); fuel consumption 27mpg (10.5l/100km).

1949-1953 TD Midget

If the YT tourer as a whole was a disappointment, fragments of it survived when Abingdon embarked on a new sports 2-seater. Some sort of Riley-MG or uprated TC had been looked at but abandoned. Instead MG traditionalists were dismayed at the introduction of the upright and taller TD compared against the old TC with its roots withering in the 1930s. The TD's development programme consisted of little more than shortening a Y-type chassis by 5in (13cm) between the wheels and reconfiguring the TC body with slab tank at the back and traditional radiator in front. It was an informal process although serious drawings were produced, a prototype made in 1948 and production began a year later. Enthusiasts disliked steel disc instead of centre-lock wire wheels, independent front suspension, which they regarded as effete and front and rear bumpers that insulted their parking skills but were essential in countries that parked by ear. The TD even abandoned the traditional MG underslung rear chassis frame and the trim and hood were designed to be weathertight and, by comparison with previous attempts, almost comfortable. Accommodating Americans was all very well but this seemed too much.

It was a hasty judgement. Had it not been for the United States market MG might have atrophied and collapsed. As it was, the TD was not only snug and reasonably secure; it was also more robust and cornered far better. Independent suspension, a stiffer chassis and lowered gearing more than made up for the increase in weight. One of the first modifications was to pierce holes in the disc wheels, ostensibly to help brake cooling. It may have done so, yet somehow it transformed the car's appearance. Eight out of ten TDs went to the US.

BODY roadster, 2 doors, 2 seats; weight 1930lb (875.5kg). ENGINE 4 cylinders, in-line; front; 66.5mm x 90mm, 1250cc; compr 7.25:1; 54.4bhp (40.6kW) @ 5200rpm; 43.5bhp (32.5kW)/l; 63.5lbft (85Nm) @ 2600rpm. ENGINE STRUCTURE duplex roller chain-driven camshaft; pushrod overhead valve; cast iron block and top half of crankcase; aluminium ribbed sump; cast iron head; coil ignition; two semi-downdraught SU carburettors; 3-bearing crankshaft; cooling pump fan thermostat. TRANSMISSION rear wheel drive; single dry plate clutch; 4-speed manual gearbox, synchromesh on 2nd, 3rd, top; hypoid final drive 5.125:1. CHASSIS DETAILS steel welded box-section side rails and tubular cross members; independent front suspension by coil spring and wishbone, live rear axle with semi-elliptic leaf springs; Luvax-Girling lever arm hydraulic dampers; Lockheed hydraulic brakes with 9in (22.9cm) (22.7cm) drums, 2-leading shoe at front; rack and pinion steering; 12.5gal (56.8l) fuel tank; 5.50 x 15 tyres; bolt-on steel disc wheels. DIMENSIONS wheelbase 94in (239cm); track front 47.4in (120cm), rear 50in (127cm); turning circle 31.3ft (9.5m); ground clearance 6in (15cm); length 145in (368cm); width 58.3in (148cm); height 56in (142cm); PERFORMANCE contemporary test maximum speed 77.2mph (124kph); 14.4mph (23kph) @ 1000rpm; 0-60mph (96kph) 21.3sec; 16.1kg/bhp (21.6kg/kW); fuel consumption 25mpg (11.3l/100km). PRICE £751.19s.2d (£751.96). PRODUCTION 29,664.

1949 George Phillips Le Mans TC

George Phillips, Fleet Street photographer and one of the founders, with Gregor Grant and John Bolster, of *Autosport* magazine, nursed passions for MGs and the Le Mans 24 Hours Race. Phillips sought help from Abingdon on how to make his TC faster for the 1947 Brighton Speed Trials, but at Lc Mans he and co-driver RM "Curly" Dryden would be up against four Talbot-Lagos, seven Delahayes, four Delages, six Aston Martins, an HRG, hordes of Simcas and two of the new Ferrari 166MMs. A tall order. Designers were beginning to understand how important a smaller frontal area and elementary aerodynamics were, so Phillips commissioned new bodywork first from Harry Lester, later of Lester MG, then Ted Goodwin. TCs were fine for amateur club racing but needed more speed on the Sarthe circuit. Le Mans turned out to be a victory for one of the astonishing high-revving beautifully made Ferraris, Culpan and Aldington were third in their Frazer-Nash, but Dryden in the MG suffered an electrical fault out on the circuit. A mechanic was despatched to instruct the driver how to effect a cure, since nobody else could work at the trackside, but Dryden, accustomed to the leisurely application of rules in British club racing gave him a lift back to the pits and was immediately disqualified. Phillips was most disgruntled but he and Eric Winterbottom had more luck in 1950, finishing a heartening 2nd in the 1½litre class to a special Jowett Jupiter, averaging 73mph (117kph) over 24 hours.

BODY sports, 2 seats. ENGINE 4 cylinders, in-line; front; 66.5mm x 90mm, 1250cc. ENGINE STRUCTURE duplex roller chain-driven camshaft; pushrod overhead valve; cast iron block and top half of crankcase; aluminium ribbed sump; cast iron head; coil ignition; two semi-downdraught SU carburettors; 3-bearing crankshaft; cooling by pump, fan, and thermostat. TRANSMISSION rear wheel drive; Borg & Beck sdp clutch; 4-speed manual gearbox; synchromesh on second, third, and top; spiral bevel final drive 5.125:1. CHASSIS DETAILS parallel steel channel-section frame, underslung at rear, with four tubular cross-bracing and boxed-in round engine and gearbox; semi-elliptic springs front and rear with shackles; Luvax lever arm hydraulic dampers; Lockheed hydraulic brakes with 9in (22.9cm) (22.7cm) drums; Bishop cam gear steering; 16gal (72.7l) fuel tank; 4.50 x 19.0 tyres; centre-lock wire wheels. DIMENSIONS wheelbase 94in (239cm); track 45in (114cm); turning circle 37ft (11.3m); ground clearance 6in (15cm); length 139.5in (354.3cm); width 56in (142cm); height 53in (135cm). PERFORMANCE fuel consumption 22mpg (12.8l/100km); 15.9mph (25.6kph) @ 1000rpm. PRODUCTION 1.

1950-1953 TD Mk II

From 1950 to 1953 the factory offered a tuned version of the TD to which C (for competition) was added as a prefix to the chassis number (later TD3). These were not to be confused with TD2s, identified by engine numbers XPAG TD2, which merely incorporated running changes including a bigger clutch. Centrepiece of the TDC conversion was the TC Stage I cylinder head modified to different phases of tune between 57bhp (42.5kW) and 61bhp (45.5kW). The suspension was stiffened at the front by means of extra, somewhat old-fashioned looking, Andrex adjustable friction-type shock absorbers. The modified cylinder head increased the compression ratio and it had bigger valves. Identifying features on Mark IIs included a small bulge on the right side of the bonnet to accommodate the bigger inlet manifold, different badges and, on some examples, chrome radiator slats. It was possible to increase output still more for racing and the XPAG could be made to give half as much power again with a supercharger. The TD/C began as a motor racing project for Dick (Richard William) Jacobs (1916-1987) of Mill Garage Essex, an MG specialist who took on and beat the benchmarking HRGs in the spring of 1950. Three works cars, FRX 941, 942 and 943 were prepared for Jacobs, Ted Lund and George Phillips to race at Silverstone, winning a worthy team prize in a drenching wet TT at Dundrod.

BODY roadster, 2 doors, 2 seats; weight 1930lb (875.5kg). ENGINE 4 cylinders, in-line; front; 66.5mm x 90mm, 1250cc; compr 8.6:1, 8.1:1 optional; between 57bhp (42.5kW) and 61bhp (45.5kW), 45.6bhp (34kW)/l and 48.8bhp (36.4kW)/l. ENGINE STRUCTURE duplex roller chain-driven camshaft; pushrod overhead valve; cast iron block and top half of crankcase; aluminium ribbed sump; cast iron head; coil ignition, manual advance-retard; two semi-downdraught 1½in H4 SU carburettors, two SU fuel pumps; Lucas Sports Coil; 3-bearing crankshaft; cooling by pump, fan, and thermostat. TRANSMISSION rear wheel drive; single dry plate clutch; 4-speed manual gearbox, synchromesh on 2nd, 3rd, top; hypoid final drive 4.875, 4.55, or 5.125:1. CHASSIS DETAILS steel welded box-section side rails and tubular cross members; independent front suspension by coil spring and wishbone, live rear axle with semi-elliptic leaf springs; Luvax-Girling lever arm hydraulic dampers, Andrex friction dampers at front; Lockheed hydraulic brakes with 9in (22.9cm) (22.7cm) drums, 2-leading shoe at front; rack and pinion steering; 12.5gal (56.8l) fuel tank; 5.50 x 15 tyres; bolt-on steel disc wheels. DIMENSIONS wheelbase 94in (239cm); track front 47.4in (120cm); rear 50in (127cm); turning circle 31.3ft (9.5m); ground clearance 6in (15cm); length 145in (368cm); width 58.3in (148cm); height 56in (142cm). PERFORMANCE maximum speed 83mph (134kph); 0-60mph (96kph) 16.5sec; between 15.4kg/bhp (20.6kg/kW) and 14.4kg/bhp (19.2kg/kW); fuel consumption 25mpg (11.3l/100km) PRICE £751.19s.2d (£751.96). PRODUCTION 1,710.

Above: TD chassis exemplified sitting on top of a car rather than inside a perimeter frame like its successor the MGA. Design principles had not changed much for 25 years.

1951 EX172 George Phillips Le Mans TD

Above: Alec Hounslow at wheel, Syd Enever third from left.

Having competed successfully in 1949 and 1950 Phillips had an automatic entry for the 1951 Le Mans, and hatched plans, along with John Thornley, sales and service manager, for a new MG based on the TD to replace his earlier TC. EX172 was allocated to a design for a streamlined body on chassis TD/C5336. Syd Enever designed it and UMG 400 became the true inspiration of the MGA. The full-width sweeping wings, low rounded nose and tapered tail were probably more shapely than strictly low-frontal-area aerodynamic. There was a lot of talk at the time of ram air effect and a small scoop on the bonnet hopefully accelerated air into the carburettors. Phillips did not like the car much. He was unaccustomed to independent front suspension and unhappy with the high driving position dictated by sitting on top of the TD chassis. Le Mans was not a success. Trouble was attributed to the quality of the fuel and despite lowering the compression ratio the engine dropped a valve and put a hole in a piston. Phillips and his co-driver Allen Rippon carried on with three cylinders but the bearings eventually ran and it retired. Recriminations followed. Phillips complained that he had not had much help from the factory and was disappointed at the preparation. For his part Thornley disapproved of his approach to driving. The car was broken up and was regarded, quite wrongly as it happened, a disappointment all round.

BODY roadster, 2 doors, 2 seats. ENGINE 4 cylinders, in-line; front; 66.5mm x 90mm, 1250cc; compr 8.1:1. ENGINE STRUCTURE duplex roller chain-driven camshaft; pushrod overhead valve; cast iron block and top half of crankcase; aluminium ribbed sump; cast iron head; coil ignition; two semi-downdraught 1½in H4 SU carburettors, two SU fuel pumps; Lucas Sports Coil; 3-bearing crankshaft; cooling pump, fan, and thermostat. TRANSMISSION rear wheel drive; single dry plate clutch; 4-speed manual gearbox, synchromesh on 2nd, 3rd, top; auto; hypoid final drive 4.875, 4.55, or 5.125:1. CHASSIS DETAILS steel welded box-section side rails and tubular cross members; independent front suspension by coil spring and wishbone, live rear axle with semi-elliptic leaf springs; Luvax-Girling lever arm hydraulic dampers, Andrex friction dampers at front; Lockheed hydraulic brakes with 9in (22.9cm) (22.7cm) drums, 2-leading shoe at front; rack and pinion steering; 12.5gal (56.8l) fuel tank; 5.50 x 15 tyres; bolt-on steel disc wheels. DIMENSIONS wheelbase 94in (239cm); track front 47.4in (120cm), rear 50in (127cm); turning circle 31.3ft (9.5m) ground clearance 6in (15cm); length 145in (368cm); width 58.3in (148cm); approx height 48in (20.3cm) (122cm). PERFORMANCE maximum speed 116mph (187kph). PRODUCTION 1.

1951-1953 YB saloon

Above: Team car. The works YBs on the 1953 RAC Rally wore Lance Corporal, Corporal and Sergeant stripes. This corporal car on show at Goodwood was driven by Len Shaw.

Improvements to the Y-type at the 1951 Earls Court Motor Show corresponded to features shared with the TD including an increase in power, although it was barely noticeable. The hypoid final drive now used throughout the Nuffield range was quieter and front brakes were changed to two-leading shoe with two cylinders for each drum, which evened out wear. Wheel sizes were reduced to 15in (38cm) instead of 16in (41cm) and the tyre section increased to 5.50. At the front an anti-roll bar was introduced and at the back heavier-duty Luvax-Girling lever arm dampers brought in as the Panhard rod was deleted. The result of these changes was a reduction in body roll. It was a pity the YB was so slow and low-geared, as its good handling could easily have accommodated more speed. Bodywork alterations to suit smaller wheels included fuller rear wings, and larger-section tyres meant an amendment to the neat spare wheel stowage under the luggage boot. Modest rally successes continued. Works entries driven by Len Shaw, Reg Holt and Geoff Holt won the team prize in the 1953 RAC Rally, Shaw finishing a keenly contested sixth overall. Dick Jacobs won his class in the Production Touring car race at Silverstone three years in a row.

BODY saloon, 4 doors, 4 seats; weight 21cwt (1067kg). ENGINE 4 cylinders, in-line; front; 66.5mm x 90mm, 1250cc; compr 7.25:1; 44bhp (34.3kW) @ 4800rpm; 35.2bhp (27.4kW)/l; 59lbft (79Nm) @ 2400rpm. ENGINE STRUCTURE duplex roller chain-driven camshaft; pushrod overhead valve; cast iron block and top half of crankcase; aluminium ribbed sump; cast iron head; coil ignition; SU 1¼in carburettor; 3-bearing crankshaft; cooling by pump, fan, and thermostat. TRANSMISSION rear wheel drive; 7¼in Borg & Beck single dry plate clutch; 4-speed manual gearbox; synchromesh on second, third, and top; Hardy Spicer needle roller bearing prop shaft; hypoid bevel final drive 5.142:1. CHASSIS DETAILS steel welded box-section side rails and tubular cross members, underslung at rear; independent front suspension by coil spring and wishbone, anti-roll bar; live rear axle with shackled semi-elliptic leaf springs; Luvax-Girling lever arm hydraulic dampers; Lockheed 2 leading-shoe (front) hydraulic brakes with 9in (22.9cm) (22.7cm) drums; rack and pinion steering; 13.5gal (61.4l) fuel tank; 5.50 x 15 tyres; bolt-on steel disc wheels, Jackall hydraulic jacking. DIMENSIONS wheelbase 99in (22.9cm) (251.5cm); track frnt 47.4in (120cm), rear 50in (127cm); lgth 161in (409cm). PERFORMANCE maximum speed 71mph (114kph); 0-60mph (96kph) 28.2sec; 24.3kg/bhp (31.1kg/kW); fuel consumption 27mpg (10.5l/100km). PRICE £671.11s 8d (£671 58) inc purchase tax. PRODUCTION 1301.

1952 EX175 MGA prototype

BMC's reluctance to invest in new model development affected morale at Abingdon. Sales of the TD were in decline and competitors such as the Triumph TR2 looked threatening. Syd Enever with his gift for practicality with inventive genius undertook a new chassis frame to deal with the problem shown up by UMG 400's high driving position. He placed the side members well apart so that the occupants could sit lower on each side of the transmission tunnel; indeed the floor was only some 6in (15.2cm) above the road. To ensure stiffness he built in a strong framework round the scuttle. Two such chassis were made, with a body added along the lines of UMG 400, which had been inspired by sports racing cars of the 1940s and 1950s such as the 1939-1940 Mille Miglia BMW 328. The shapely result, which John Thornley memorably called "UMG400 put right", was a prototype, HMO 6, with hood, sidescreens, bumpers, and although the tall T-series engine demanded a bulge in the bonnet top, it was adopted almost without alteration for the MGA. Slightly bigger than EX172, the stimulating EX175 was, alas, put immediately on ice. Leonard Lord had just confirmed at the Earls Court Motor Show that he had bought into the Austin-Healey 100, which meant that in October 1952 he turned EX175 down. In 1953 Enever streamlined the second EX175 with wheel spats and a bubble cockpit to see if it could be used as a record-breaker, but it proved unsuitable in wind-tunnel tests. The scheme was abandoned until re-incorporated in EX179 three years later.

BODY roadster, 2 doors, 2 seats; weight approx 1930lb (875.5kg). ENGINE 4 cylinders, in-line; front; 66.5mm x 90mm, 1250cc; cr 7.25:1; 54.4bhp (40.6kW) @ 5200rpm; 43.5bhp (32.5kW)/l; 63.5lbft (85Nm) @ 2600rpm. ENGINE STRUCTURE duplex roller chain-driven camshaft; pushrod overhead valve; cast iron block and top half of crankcase; aluminium ribbed sump; cast iron head; coil ignition; two semi-downdraught 1½in H4 SU carburettors, two SU fuel pumps; Lucas Sports Coil; 3-bearing crankshaft; cooling pump, fan, and thermostat. TRANSMISSION rear wheel drive; single dry plate clutch; 4-speed manual gearbox, synchromesh on 2nd, 3rd, top; auto; hypoid final drive 4.875, 4.55, or 5.125:1. CHASSIS DETAILS steel welded box-section perimeter side rails and tubular cross members; independent front suspension by coil spring and wishbone, live rear axle with semi-elliptic leaf springs; Luvax-Girling lever arm hydraulic dampers; Lockheed hydraulic brakes with 9in (22.9cm) (22.9cm) drums, 2-leading shoe at front; rack and pinion steering; 12.5gal (56.8l) fuel tank; 5.50 x 15 tyres; bolt-on steel disc wheels. DIMENSIONS wheelbase 94in (239cm); track front 47.4in (120cm), rear 50in (127cm); turning circle 31.3ft (9.5m); grnd clearance 6in (15cm); length 145in (368cm); width 58.3in (148cm); height 56in (142cm). PERFORMANCE 16.1kg/bhp (21.6kg/kW). PRODUCTION 2 chassis frames, 1 body.

1953-1954 TF Midget 1250

The raked grille, lowered bonnet, faired-in headlamps and new interior of the TF were not at first highly regarded at Abingdon. It was looked on as something of a stop-gap and one that not even its creators cared for much. The factory convinced itself, not without cause, that it had been left behind in the race to modernity exemplified by rivals. The TF's radiator was now imitation, with a dummy filler cap and header tank under the bonnet and, mindful of criticism of disc wheels on the TD, new hubs were made so that wire wheels could be offered as an option. The engine was TD Mark II, otherwise not much had changed. The TF was not even well received on its introduction at the 1953 motor show. Press and enthusiasts alike had become mesmerised by what they thought more "streamlined" Triumph TR2s and Austin-Healeys that did 100mph (160.9kph) for about the price of a TF. Jaguar was making cars that did more than 130mph (209kph) so the little MG was well outclassed. It was years before the TF's elegant proportions and traditional structure came to be regarded as classic. New model development was still being done at Abingdon much as it had been 30 years before.

No drawings were made for the new shape; Freddie Wake and Bert Wirdnam hammered out the flowing wings in a fortnight. A panel beater who worked with them, Billy Wilkins, had worked at Raworth, the firm that built some of Kimber's first cars and this TF, in some respects, was scarcely different.

BODY roadster, 2 doors, 2 seats; weight 1930lb (875.5kg). ENGINE 4 cylinders, in-line; front; 66.5mm x 90mm, 1250cc; compr 8.0:1; 57bhp (42.5kW) @ 5500rpm; 45.6bhp (34kW)/l; 65lbft (87Nm) @ 3000rpm. ENGINE STRUCTURE duplex roller chain-driven cam; pushrod ohv; cast iron block and top half of crankcase; aluminium ribbed sump; cast iron head with bathtub combustion chambers; coil ignition; two semi-downdraught SU carbs; 3-bearing crank; cooling by pump, fan, and thermostat. TRANSMISSION rear wheel drive; 8in (20.3cm) sdp clutch; 4-speed manual gearbox, synchromesh on 2nd, 3rd, top; auto; hypoid final drive 4.875:1. CHASSIS DETAILS steel welded box-section side rails, tubular cross members, and scuttle reinforcement; ifs by coil spring and wishbone, live rear axle with semi-elliptic leaf springs; Luvax-Girling lever arm hydraulic dampers; Lockheed hydraulic brakes with 9in (22.9cm) (22.9cm) drums, 2-leading shoe at front; rack & pinion steering; 12.5gal (56.8l); 5.5 x 15 tyres; bolt-on steel discs, wire wheels optional. DIMENSIONS wheelbase 94in (239cm); track 47.4in (120cm) front, 50in (127cm) rear, 48.4in (123cm)/50.8in (20.3cm) (129cm) with wire wheels; turning circle 31.3ft (9.5m); ground clearance 6in (15cm); length 147in (17.8cm) (373cm); width 58.3in (148cm); height 54.5in (138cm). PERFORMANCE maximum speed 73.5mph (118kph); 15.3mph (24.5kph) @ 1000rpm; 0-60mph (96kph) 18.9sec; 15.4kg/bhp (20.6kg/kW); 25mpg (11.3l/100km). PRICE £780.5s.10d (£780.29). PRODUCTION 6,200.

1953 Arnolt, Ghia-Aigle, and Shipsides

Above: Arnolt MGs. Only 65 coupes and 35 roadsters were made and it was several years before the last was sold.

The shorter TD proved a better basis for coachbuilders than the YA. Nuccio Bertone's struggling Carrozzeria showed two at the 1952 Turin Motor Show through Nuffield distributors Fattori and Mantani of Rome. Open and closed versions were in the fashionable full-width Italian style with an MG radiator shell upright at the front.

Pretty cars with good proportions, they caught the attention of Chicago MG importer Harold 'Wacky' Arnolt, who also imported Aston Martins, Bentleys and Bristols. He felt sure he could sell more TDs if they looked like this, and without the quirkiness or frailty of Italian cars. He ordered 100 of each sort from Bertone and was immediately appointed a vice-president. They did not sell quickly however. New York importer Inskip tried something the same with TDs stretched by 10in (25.4cm) (25.4cm) to provide four seats and bigger doors. At $2925, only 12 were made. Ghia-Aigle made six specially-bodied TDs for Swiss importer JH Keller. They too featured upright MG radiator shells and wide bodywork. The open cars had tail-fins rather like the later Facellia, and MG took a keen interest in the result. Nottingham dealer Shipsides commissioned a rather bulbous coupe as an experiment but only one was ever built.

Spec Arnolt 1¼litre: BODY coupe, convertible; 2 doors, 2+2 seats; weight 2103lb (954kg) coupe, 2083lb (945kg) convertible. ENGINE 4 cylinders, in-line; front; 66.5mm x 90mm, 1250cc; compr 7.25:1; 54.4bhp (40.6kW) @ 5200rpm; 43.5bhp (32.5kW/)/l; 63.5lbft (85Nm) @ 2600rpm. ENGINE STRUCTURE duplex roller chain-driven camshaft; pushrod overhead valve; cast iron block and top half of crankcase; aluminium ribbed sump; cast iron head; coil ignition; two semi-downdraught SU carburettors; 3-bearing crankshaft; cooling by pump, fan, and thermostat. TRANSMISSION rear wheel drive; Borg and Beck sdp clutch; 4-speed manual gearbox, synchromesh on 2nd, 3rd, top; hypoid final drive 5.125:1. CHASSIS DETAILS steel welded box-section side rails and tubular cross members; independent front suspension by coil spring and wishbone, live rear axle with semi-elliptic leaf springs; Luvax-Girling lever arm hydraulic dampers; Lockheed hydraulic brakes with 9in (22.9cm) (22.9cm) drums, 2-leading shoe at front; rack and pinion steering; 12.5gal (56.8l) fuel tank; 5.50 x 15 tyres; bolt-on steel disc wheels, Borrani wires on Bertone Arnolt TD. DIMENSIONS wheelbase 94in (239cm); track front 47.4in (120cm), rear 50in (127cm); turning circle 31.3ft (9.5m); ground clearance 6in (15cm); length 160in (406cm); width 60in (152cm); height 55in (140cm). PERFORMANCE 17.5kg/bhp (23.5kg/kW); 14.4mph (23.2kph) @ 1000rpm; fuel consumption 25mpg (11.3l/100km). PRICE US POE $2995 rising to $3585. PRODUCTION 100 Arnolt MGs.

1953-1956 Magnette ZA

The furore over the MG TF in October 1953 was nothing compared with the cynicism surrounding the launch, at the same motor show, of the Magnette. There was indignation about the adoption of an old MG name for what seemed a clone of the Wolseley 4/44 introduced the previous year. The rationale for the fussing was the more obscure since the Wolseley had effectively an MG engine, the 1250cc XPAG, and the Magnette a BMC B-series engine descended from an Austin. It looked like BMC opportunism, yet both cars were conceived at the same time. The Magnette was delayed to allow time for the B-series, a fruit of the Nuffield-BMC merger, to be engineered into it. As so often with MGs the excellent qualities of the Magnette, designed by Gerald Palmer, became more apparent later, although Marcus Chambers was critical in their early competition efforts owing to their poor traction; "tail-happy" was his description although in reality they were a lot better than many contemporaries.

It was salutary that MG, for so long seen as a marque in its own right, was now being disparaged through its long-established association with volume manufacture. Certainly not much subtlety was shown about the distinguishing features between the Magnette and the Wolseley, unlike those between the Y-type and its Morris clone, the Series E. The Magnette's monocoque body was 2in lower than the Wolseley's and the differences extended to the floor pan, which had to accommodate a stick gearshift, some body pressings, different lights, a plush interior and styling details such as the clever chrome finisher on the front wheelarch. The Magnette felt more sporting, the Wolseley had a steering column change, and although both handled well the Magnette was superior in almost every way.

BODY saloon, 4 doors, 4 seats; weight 2507lb (1137kg). ENGINE 4 cylinders, in-line; front; 73mm x 89mm, 1489cc; compr 7.15:1; 60bhp (44.7kW) @ 4600rpm; 40.3bhp (30kW)/l; 76.1lbft (102Nm) @ 3000rpm. ENGINE STRUCTURE pushrod overhead valve; chain-driven camshaft; cast iron cylinder head and block; 2 SU inclined carburettors, 12-volt coil ignition; SU electric fuel pump; 3-bearing crankshaft; engine rubber-mounted. TRANSMISSION rear wheel drive; 8in (20.3cm) single dry plate clutch; 4-speed manual gearbox, synchromesh; Hardy Spicer prop shaft; hypoid bevel final drive 4.875:1. CHASSIS steel monocoque structure; ifs by coil spring and wishbone; live back axle, semi-elliptic springs; telescopic dampers; hydraulic 10in (25.4cm) (25.4cm) drum brakes, 2LS at front; rack and pinion steering; 9.5gal (43.2l) fuel tank; 5.50-15 tyres, steel disc wheels. DIMENSIONS wheelbase 102in (259cm); track 51in (129.5cm); turning circle 37ft (11.3m); ground clearance 7in (17.8cm) (18cm); length 169in (22.9cm) (429cm); width 63in (160cm); height 58in (20.3cm) (147cm). PERFORMANCE maximum speed 79.5mph (128kph); 15.3mph (24.5kph) @ 1000rpm; 0-60mph (96kph) 22.6sec; 19kg/bhp (25.4kg/kW); fuel consumption 25.3mpg (11.2l/100km). PRICE £914 17s 6d (£914.87). PRODUCTION 12,754.

1954-1955 TF 1500

Left: TF 1500 brochure artist's drawings of cars (with riding mechanics) somehow omitted the "1500" label on the bonnet sides.

The last traditional MG was made with wooden formers under the bodywork, a proper Auster windscreen that folded flat, scuttle cowls and doors hinged at the rear. Getting in and out easily was still thought more important than any risk of leg injuries. Private owners had already been boring out XPAG 1250cc engines to extract more power, a chancy business given the thin cylinder walls. The factory re-cored some engines to allow 72mm to be bored out safely, but these were not put into production. In August 1954 EX179 ran with such an engine at Utah, so the opportunity was taken to put it into production for a 1½litre TF. Sufficient metal for the larger bores was achieved by doing away with the water spaces between cylinders 1 and 2, and 3 and 4. The bigger bore brought a proportional increase in compression ratio. The only sign of the larger engine was a discreet "TF" 1500 badge on the bonnet side panels; there was never any explanation for the quotation marks.

Performance was improved, though at about 85mph it still ran up against an impenetrable aerodynamic wall of a 16sq ft (1.49sq m) frontal area. The TF 1500 was never going to be as fast as its Triumph or Austin-Healey contemporaries, but it was cherished by collector and connoisseurs as one of the last survivors, along with Morgan, of a dying strain of sports car structures. With production starting in July 1954 it bought Abingdon some valuable time before the introduction of the MGA.

BODY roadster, 2 doors, 2 seats; weight 1930lb (875.5kg). ENGINE 4 cylinders, in-line; front; 66.5mm x 90mm, 1466cc; compr 8.3:1; 63bhp (47kW) @ 5000rpm; 43bhp (32kW)/l; 65lbft (87Nm) @ 3000rpm. ENGINE STRUCTURE duplex roller chain-driven camshaft; pushrod overhead valve; cast iron block and top half of crankcase; aluminium ribbed sump; cast iron head; coil ignition; two semi-downdraught SU carburettors; 3-bearing crankshaft; cooling by pump, fan, and thermostat.TRANSMISSION rear wheel drive; 8in (20.3cm) single dry plate clutch; 4-speed manual gearbox, synchromesh on 2nd, 3rd, top; hypoid final drive 4.875:1. CHASSIS DETAILS steel welded box-section side rails tubular cross members and scuttle reinforcement; independent front suspension by coil spring and wishbone, live rear axle with semi-elliptic leaf springs; Luvax-Girling lever arm hydraulic dampers; Lockheed hydraulic brakes with 9in (22.9cm) (22.9cm) drums, 2-leading shoe at front; rack and pinion steering; 12.5gal (56.8l) fuel tank; 5.50 x 15 tyres; bolt-on steel disc wheels, wire wheels optional. DIMENSIONS wheelbase 94in (239cm); track front 47.4in (120cm) rear 50in (127cm), or 48.4in (123cm) and 50.8in (20.3cm) (129cm) with wire wheels; turning circle 31.3ft (9.5m); ground clearance 6in (15.2cm); length 147in (17.8cm) (373cm); width 58.3in (148cm); height 54.5in (138cm). PERFORMANCE maximum speed 88mph (142kph); 15.2mph (24.5kph) @ 1000rpm; 0-60mph (96kph) 16.3sec; 13.9kg/bhp (18.6kg/kW); fuel consumption 25mpg (11.3l/100km). PRICE £780 5s 10d (£780.29). PRODUCTION 3400.

1954 EX179

George Eyston not only saw the publicity value of record-breaking, but as a director of Castrol he was also in a position to do something about it. He had disagreed with Kimber's sales of The Magic Midget to Germany in 1934 but with EX135, Goldie Gardner's car, effectively out of MG's control he secured Castrol sponsorship for a new record-breaker. This was based on the redundant chassis of the MGA prototype EX175 and the approaching TF 1500's larger-bore (72mm instead of 66.5mm) XPAG (by now XPEG) engine of 1466cc. Power was increased to 84bhp (62.6kW) at 6000rpm with a compression ratio of 9.3:1 and it ran on ordinary pump fuel. Wilson McComb described it picturesquely as "an extraordinarily inoffensive unit to use in a record-breaker." It was his way of saying it was a bit weak.

Not unlike the original Gardner car in appearance, Eyston and the American Ken Miles took eight Class F records at 153.69mph (246.72kph) and averaged over 120mph (192.6kph) for 12 hours at Bonneville. In 1956 the car was taken back to the salt flats with an unsupercharged 1489cc Twin Cam engine and, driven again by Miles and former racing motorcyclist Johnny Lockett, it took 16 records at up to 170.15mph (273.14kph). McComb would have approved of that.

With a 948cc A-series engine from the yet to be announced Austin-Healey Sprite (and later MG Midget), giving 57bhp (42.5kW) it broke three international and 25 American standing-start records driven by Tommy Wisdom and David Ash, author of Automobile Almanac. An Austin-Healey development Sprite was found aerodynamically unsuitable so EX179 was redesignated EX219 and became officially an Austin-Healey for the occasion. With Phil Hill at the wheel and the engine supercharged, this versatile car then did the flying mile at 143.47mph (230.31kph). A confident British Motor Corporation was thus able to advertise its new cars proudly, and perfectly truthfully, with a "record-breaking" engine.

EX179 was a vindication of Enever's pragmatism and ingenious low-built chassis, although the car was mainly the development work of Terry Mitchell, MG's chief chassis designer. It was one of the first tasks he undertook after moving from Cowley to the design office re-established at Abingdon. Enever caused Mitchell to retain the characteristics of the Gardner MG, and the bubble canopy was moved from left to right hand drive to suit the regulations. The aluminium body was made by Midland Sheet Metal.

Further vindication of the body design came with a series of tests carried out by clever technician Michael Scarlett of *Autocar* in 1982, with drag coefficients of some notable cars measured in the MIRA wind tunnel. These included the D-type Jaguar (Cd 0.49), Lotus Elite (Cd 0.35) and Scarlett compared them with those obtained elsewhere. It was always difficult to get a road car Cd much under the Elite's 0.35. A Porsche 924 of 0.37 was representative, a Citroen GSA's 0.34 was exceptional and the 0.26 of EX179 remarkable. It might have been even better had there been an undertray between the back of the nose and the toeboard.

In other words the streamlining on top was amazingly good but the underside was neglected. Designers invariably failed to appreciate the importance underneath, where the slipstream was squeezed between a jumble of components at speed and the stationary road, so it was much misunderstood. The wind tunnel also showed more lift than expected for a 170mph (273kph) car but the figure that mattered, the Cd X frontal area of 3.0, was by far the lowest (the Jaguar was 6.32, the Lotus 5.26) and in a test to determine how the airflow behaved over the body, the only turbulence proved to be immediately behind the cockpit. Scarlett described the way the airflow clung to the entire length of the MG as "a marvel". EX179 was honourably retired to the British Motor Industry Heritage Trust (later the British Motor Museum) at Gaydon (below).

BODY single-seat; enclosed; weight 2198lb (997kg) including driver and 30gal (136.4l) fuel. ENGINE various 4 cylinders, in-line; front, see text. TRANSMISSION rear wheel drive; sdp clutch; 4-speed manual gearbox, synchromesh on 2nd, 3rd, top; hypoid final drive 4.875, 4.55, 5.125:1 or special gearing for records. CHASSIS DETAILS steel welded square-section perforated tubing side rails and tubular cross members; ifs by coil spring and wishbone, live rear axle with semi-elliptic leaf springs; Luvax-Girling lever arm hydraulic dampers; Lockheed hydraulic brakes rear wheels only, with 9in (22.9cm) (22.9cm) drums, 2-leading and trailing shoe; rack and pinion steering; 12.5gal (56.8l) tank; 5.50 x 15 tyres; bolt-on steel disc wheels. DIMENSIONS wheelbase 95in (241cm); track front 47.4in (120cm), rear 50in (127cm); turning circle 31.3ft (9.5m); length 188.9in (22.9cm) (480cm); width 63.5in (161.3cm); height to cockpit top 40.4in (102.6cm). PERFORMANCE see text. PRODUCTION 1.

Below: EX 179 in British Motor Museum Gaydon

1955 EX182 Le Mans team cars

Above: One of the Le Mans cars racing at the 1955 TT, Ted Lund at the wheel.

The Le Mans MGs of 1955 displayed, with their perimeter frame and maturity, welcome hallmarks of a production-ready car. Under the direction of Alec Hounslow, Nuvolari's riding mechanic in the historic TT of 1933, they were prototypes in the spirit of regulations that had been drawn up in the 1920s. These had developed over the years with the aim of encouraging manufacturers to try out new models in the full glare of publicity. This project, known as EX182, resulted in four production lookalikes with 18-gauge aluminium bodies and Weslake-developed cylinder heads. They were truly harbingers of things to come, exquisitely proportioned, in every sense a sports car classic and one of the most widely admired MGs ever. The bodies had the requisite seats and doors, the passenger side covered with a fairing for the race and it was a shame that such a brave initiative was overwhelmed by events.

The 1955 race was a disaster; a Mercedes-Benz 300SLR crashed into the crowd, killing its driver Pierre Levegh and over 80 spectators. One of the MGs crashed almost at the same time, badly injuring Dick Jacobs, a staunch MG supporter whose connections with the company went back to 1937. The race went on. Abandonment would only have created further mayhem and the MGs acquitted themselves nobly, finishing 5th and 6th in the class behind three Porsches and an OSCA. They averaged 86.17mph (138.67kph) and 81.97mph (131.9kph), finishing 12th and 17th overall while the Triumph TR2s that had prompted BMC to set up its competitions committee were 14th, 15th and 19th. MG had no better fortune in September at the Ulster TT, also blemished by fatal accidents and leading to LP Lord's pressure to concentrate on rallying. MGs in the TT had available twin overhead camshaft engines that foresaw development of the later MGA Twin Cam. Magnette mastermind Gerald Palmer drew up plans for a twin cam based on the B-series, with 90deg valves, made by Morris at Coventry but it was 1958 before a production one was available to try in an MGA. The Dundrod car's was a prototype. Austin at Longbridge planned a 66deg twin cam of its own that ran at Le Mans and while smooth it was never made ready for the TT and was subsequently abandoned.

BODY sports, 2 seats; weight 1596lb (724kg). ENGINE 4 cylinders, in-line; front; 73.025mm x 88.9mm, 1489cc; compr 9.4:1; 82.5bhp (61.5kW) @ 6000rpm; 55.3bhp (41.3kW)/l; 85lbft (114Nm) @ 4500rpm. ENGINE STRUCTURE pushrod ohv; chain-driven camshaft; cast iron cylinder head and block, no gasket; two 1¾in SU inclined carburettors, 12-volt coil ignition; two SU electric fuel pumps; 3-bearing crank with lead-bronze shell bearings. TRANSMISSION rear wheel drive; Borg & Beck sdp 8in (20.3cm) clutch; 4-speed manual close-ratio gearbox, synchromesh 2nd, 3rd, top; hypoid final drive with bevel-type differential 3.7:1. CHASSIS DETAILS steel perimeter frame with scuttle reinforcement; ifs by coil spring and unequal length wishbone; live rear axle with semi-elliptic springs; Armstrong lever arm dampers front and rear; Lockheed hydraulic brakes 10in (25.4cm) (25.4cm) drums, 2LS at front; rack and pinion steering; 20gal (90.9l) fuel tank; front 5.60–15tyres, rear 6.00-15; centre-lock wire wheels. DIMENSIONS wheelbase 94in (239cm); track front 47.5in (127cm), rear 48.8in (20.3cm) (124cm); turning circle 28ft (8.5m); ground clearance 6in (15cm); length 150in (381cm); width 58in (20.3cm) (147cm); height 41in (104cm). PERFORMANCE maximum speed 103mph (165.8kph) (with windscreen and 4.3 axle), 119.5mph (192.3kph) at Le Mans with metal cover on passenger seat and racing screen; 21mph (33.8kph) @ 1000rpm; 0-60mph (96kph) 13.8sec; 8.8kg/bhp (11.8kg/kW); fuel consumption 23.8mpg (11.9l/100km). PRODUCTION 4.

Above: Replica at a Goodwood Revival missing the distinctive Lucas Flamethrower within the grille, with the Morris-Commercial "Comps" transporter made for the 1955 team cars.

1955-1959 MGA

Above: MG historian Wilson McComb feels at home during his spell of duty in the Abingdon press office.

If Abingdon's plans had worked, the Le Mans cars would have been even closer to production MGAs, announced almost at the same time. Pressed Steel wanted to stamp out the bodies for the new car with what were known as soft dies. These were cheap and for relatively small production runs should have been perfectly adequate. In the event they were nothing of the sort. The plastic dies broke after only a few applications and the whole MGA production schedule was delayed by months. The A appellation was applied because with the ZA and ZB Magnettes the management felt rather curiously that the end of the alphabet had been reached. It was supposed to represent a fresh start for MG, so it was launched as the MGA and the announcement, which was intended to be in the spring of 1955, was put off until the autumn. By then the clever perimeter frame into which Enever had built such strength, the B-series

BMC engine and the well-proportioned style were no longer much of a mystery. All that remained was to confirm that the ride and handling were up to scratch and, although not as fast as a TR2 or Austin-Healey 100, the MG pedigree and its superior handling carried the day. More than 13,000 were produced during the first year, far exceeding the total production of the TC in the four post-war years it was made. The MGA was off to a flying start.

BODY sports, 2 doors, 2 seats; weight 1988lb (902kg). ENGINE 4 cylinders, in-line; front; 73.025mm x 88.9mm, 1489cc; compr 8.3:1; 68bhp (50.7kW) @ 5500rpm; 45.7bhp (34kW)/l; 77.4lbft (104Nm) @ 3500rpm. ENGINE STRUCTURE pushrod overhead valve; chain-driven camshaft; cast iron cylinder head and block; 2 SU inclined carburettors, 12-volt coil ignition; SU electric fuel pump; 3-bearing crankshaft; engine rubber-mounted. TRANSMISSION rear wheel drive; Borg & Beck sdp 8in (20.3cm) clutch; 4-speed manual gearbox, synchromesh 2nd, 3rd, top; hypoid bevel final drive 4.3:1; 4.55 optional. CHASSIS DETAILS steel perimeter frame with scuttle reinforcement; ifs by coil spring and unequal length wishbone; live rear axle with semi-elliptic springs; Armstrong lever arm dampers front and rear; Lockheed hydraulic brakes 10in (25.4cm) (25.4cm) drums, 2LS at front; rack and pinion steering; 10gal (45.5l) tank; 5.60–15; centre-lock wire wheels optional. DIMENSIONS wheelbase 94in (239cm); track front 47.5in (121cm), rear 48.8in (20.3cm) (124cm); turning circle 28ft (8.5m); ground clearance 6in (15cm); length 156in (396cm); width 58in (20.3cm) (147cm); height 50in (127cm). PERFORMANCE maximum speed 98mph (157.7kph); 17mph (27.4kph) @ 1000rpm; 0-60mph (96kph) 15.6sec; 13.3kg/bhp (17.8kg/kW); fuel con 27mpg (10.5l/100km). PRICE £844 0s 10d (£844.04). PRODUCTION 58,750.

Above: Exquisite proportions. The author's MGA.

1956-1959 MGA Coupe

Many aspiring sports 2-seater drivers disliked the draughtiness of soft-tops and sidescreens in the British climate, so proprietary detachable hard-tops proliferated. An aluminium one from coachbuilders Vanden Plas was used on the Alpine Rally cars in July. However soon after its launch the MGA's engine power was increased slightly and late in 1956 came the closed coupe, its 117lb (53kg) of extra weight more than compensated for by better aerodynamics. Historically there were precedents in MG's tradition of salonettes and contemporaneously Jaguar was enjoying huge success with the stylish two-seat XK 120 Coupe. The MG, with its deeply curved front and rear windows, was an immediate success, its huge glass area banishing any feelings of claustrophobia in the compact interior. The enhanced stiffness of the body even improved the handling. Detailing was exemplary, from the neat upward-pull door handles to the well carpeted and upholstered interior. However, it was noisy inside and there were difficulties in making the wind-up windows truly draught proof. Reputedly the coupe was designed round the curved back window of the A55 Austin Cambridge, which fitted precisely in place of the detachable windscreen of the MGA sports. With the BMC parts bin able to supply the glass, no additional outside supplier was needed for the relatively small number required.

BODY sports coupe, 2 doors, 2 seats; weight 2105lb (955kg). ENGINE 4 cylinders, in-line; front; 73.025mm x 88.9mm, 1489cc; compr 8.3:1; 72bhp (53.7kW) @ 5500rpm; 48.4bhp (36.1kW)/l; 77.4lbft (104Nm) @ 3500rpm. ENGINE STRUCTURE pushrod ohv; chain-driven cam; cast iron cyl head and block; 2 SU inclined carbs, 12-volt coil ignition; SU electric fuel pump; 3-brg crankshaft; engine rubber-mounted. TRANSMISSION rear wheel drive; Borg & Beck sdp 8in (20.3cm) clutch; 4-speed manual gearbox, synchromesh 2nd, 3rd, top; hypoid bevel final drive 4.3:1; 4.55 optional. CHASSIS DETAILS steel perimeter frame w scuttle reinforcement ifs by coil spring and unequal length wishbone; live rear axle with semi-elliptic springs; Armstrong lever arm dampers front and rear; Lockheed hydraulic brakes 10in (25.4cm) (25.4cm) drums, 2LS at front; rack and pinion steering; 10gal (45.5l) tank; 5.60–15; centre-lock wire wheels optional. DIMENSIONS wheelbase 94in (239cm); track 47.5in (121cm), rear 48.8in (20.3cm) (124cm); turning circle 28ft (8.5m); ground clearance 6in (15cm); length 156in (396cm); width 58in (20.3cm) (147cm); height 50in (127cm). PERFORMANCE maximum speed 100.5mph (161.7kph); 17mph (27.4kph) @ 1000rpm; 0-60mph (96kph) 15sec; 13.3kg/bhp (17.8kg/kW); 27mpg (10.5l/100km). PRICE £1087. 7s (£1087.35). PRODUCTION 58,750 (all MGA 1500).

1956-1958 Magnette ZB

Against Standard Ensigns, Ford Consuls, Hillman Minxes, and Morris Oxfords the Magnette's crisp handling and good proportions had pure MG qualities.

The second Magnette had a more powerful engine and the axle ratio was raised for quieter cruising. Some aspects of production were still somewhat archaic. The big wrap-round back window had to be hand-cut on finished bodies made on the Pressed Steel line at Cowley. Two-colour paintwork was known as Varitone although quite a lot of ZB Magnettes were soberly finished in a single colour.

The ZB was distinguished in its top speed of nearly 90mph, precise rack and pinion steering, cornering poise and a calm air of well-being and good taste. "The air of quality that follows from the use of first-class hide upholstery and pleasingly grained wood for the facia panel and door cappings is immediately apparent. It is matched by small details … well-positioned pedals … generous parcel tray… and high quality carpets firmly fixed to the floor," said *The Motor*. The re-established competitions department at Abingdon under Marcus Chambers tentatively entered Magnettes in rallies and they won their class in the 1958 BRSCC saloon car championship. Chambers wanted to develop the model further with the Twin-Cam engine. It handled well and its traction problem would have had to be addressed, but it was too heavy with such modest power so its sporting career was brief. Manumatic transmission, with the clutch operated by an electrical switch in the gear lever, was no more of a success in the Magnette than it was anywhere else.

BODY saloon, 4 doors, 4 seats; weight 2507lb (1137kg). ENGINE 4 cylinders, in line; front; 73mm x 89mm, 1489cc; compr 8.3:1; 68bhp (50.7kW) @ 5200rpm; 45.7bhp (34kW)/l; 831lbft (111Nm) @ 3000rpm. ENGINE STRUCTURE pushrod overhead valve; chain-driven camshaft; cast iron cylinder head and block; 2 SU inclined carburettors, 12-volt coil ignition, vacuum timing control; SU electric fuel pump; 3-bearing crankshaft; engine rubber-mounted.TRANSMISSION rear wheel drive; 8in (20.3cm) single dry plate clutch; 4-speed manual gearbox, synchromesh; Hardy Spicer prop shaft; hypoid bevel final drive 4.55:1. CHASSIS DETAILS steel monocoque structure; ifs by coil spring and wishbone; live back axle, semi-elliptic springs; telescopic dampers; hydraulic 10in (25.4cm) (25.4cm) drum brakes, 2LS at front; rack and pinion steering; 9.5gal (43.2l) fuel tank; 5.50-15 tyres, steel disc wheels. DIMENSIONS wheelbase 102in (259cm); track 51in (129.5cm); turning circle 37ft (11.3m); ground clearance 7in (17.8cm) (17.8cm); length 169in (22.9cm) (429.3cm); width 63in (160cm); height 58in (20.3cm) (147.3cm). PERFORMANCE maximum speed 85.8mph (138kph); 16.4mph (26.4kph) @ 1000rpm; 0-60mph (96kph) 20.8sec; 16.7kg/bhp (22.4kg/kW); fuel consumption 25.5mpg (11.1l/100km). PRICE £1072 7s (£1072.35). PRODUCTION 23,846.

1957-1959 EX181

Syd Enever decided that the last of the Abingdon record-breakers should be a classic teardrop, like an aircraft drop-tank with cutouts for the wheels. Alleged to give 30 percent less drag than EX179, and with Abingdon's tradition for irreverent nicknames, it was soon known as the Roaring Raindrop. Following the successful centre-seat Cooper-Bristol in the mid-1950s, prejudice against engines behind the driver (a legacy of supposedly "difficult" Auto Unions) was diminishing. Jack Brabham's 1955 Cooper-Bristol that competed in the British Grand Prix confirmed that the theoretical advantages of a low polar moment of inertia were true and EX181's engine was duly located aft.

The chassis was a tubular affair, the engine a supercharged 1489cc Twin Cam, and the car appeared with and without the tail-fin that Eyston advocated. The goal in 1957 was 250mph (400kph) but the team would have been content with 240mph, or four miles a minute. Stirling Moss, five days after winning a famous victory at Pescara for Vanwall, achieved 245.64mph (394.33kph) on damp salt. It was 50mph faster than he had ever driven in his life and he described the experience many years later as quite unnerving, particularly when he found he could not use the gearbox to slow. American world champion Phil Hill, who began his career in a TC MG, set EX181's 1959 records using a 1506cc version of the engine for Class E. It was now delivering 300bhp (223.7kW), or some three times its standard output. It was the last occasion the car was driven competitively.

BODY single-seater record breaker; dry weight 1655lb (751kg). ENGINE 4 cylinders, in-line; front; 75.4mm x 89mm, 1588cc, later 1506cc; 290bhp (216.3kW) @ 7300rpm; 182.6bhp (136.2kW)/l. ENGINE STRUCTURE two chain-driven overhead camshafts; cast iron block, aluminium head; finned aluminium sump; twin 3/16in SU carbs; 3-bearing crank; Shorrock supercharger running at 32psi (2.206bar) @ 7000rpm. TRANSMISSION rear wheel drive; 7¾ 3-plate clutch; 4-speed manual gearbox; final drive 1.94:1 and 1.825:1. CHASSIS DETAILS tubular steel structure; ifs by coil spring and unequal length wishbone; De Dion rear with quarter elliptic springs; single hydraulic disc brake on half shaft; rack and pinion steering; 7.9gal (36l) tank; Dunlop ultra-high speed tyres on 15in steel disc wheels. DIMENSIONS wheelbase 96in (244cm); length 181.5in (461cm); width 64.3in (163cm); height 38.3in (97cm). PERFORMANCE maximum speed 254.91mph (409.2kph); 2.6kg/bhp (3.5kg/kW). PRODUCTION 1.

Left: Stirling Moss eases himself into the cramped cockpit of a car he described as one of the most uncomfortable he had ever driven.

1957-1959 EX186 proposed Le Mans car

Above: MG historian David Knowles discovered that John Thornley and Syd Enever had been behind the project. It was only around Christmas 1960 that an order given to get rid of it. Kjell Qvale sold it on some years later to a Californian enthusiast.

A continuing assault on Le Mans was planned after the 1955 race. Preparations were already in hand for a more radical design even beforehand. Drawings were done and a model made before the racing programme was abandoned following the disaster. Work on EX186 only ceased in 1959 when a sole example was built *sub rosa* to revive a return to Le Mans. Once BMC found out it was disposed of to Kjell Qvale, the San Francisco importer and so far as the management at the top was concerned, disremembered. The strategy of chassis designer Terry Mitchell was to have a Twin Cam engine in a much-modified MGA chassis, with De Dion rear suspension. The body was the work of Denis Williams, based on typical long-distance endurance cars of the time, with a low full width windscreen and head fairing and made by Midland Sheet Metal. It was never certain how many cars were proposed, perhaps as many as ten might have been laid down had MG continued racing. Ted Lund tried to get it released and run as a private entry at Le Mans but was unsuccessful in the face of BMC's intransigence and it remained another of MG's lost initiatives.

BODY open, 2 seats. ENGINE 4 cylinders, front; 75.4mm x 89mm, 1588cc; compr 9.9:1 (later reduced to 8.3); 108bhp (80.5kW) @ 6700rpm; 68bhp (50.7kW)/l; 104lbft (140Nm) @ 4500rpm. ENGINE STRUCTURE 2 chain-driven overhead camshafts; cast iron block, aluminium cylinder head; finned aluminium sump 2 SU H6 carburettors, 12-volt coil ignition; SU electric fuel pump; 3-bearing crankshaft; engine rubber-mounted. TRANSMISSION rear wheel drive; Borg & Beck sdp 8in (20.3cm) clutch; 4-speed manual gearbox, synchromesh 2nd, 3rd, top; Hardy Spicer prop shaft; hypoid bevel final drive 4.3:1. CHASSIS DETAILS tubular steel frame; ifs by coil spring and unequal length wishbone; De Dion rear axle; Armstrong lever arm dampers front and rear; Dunlop disc brakes; Cam Gears rack and pinion steering; 10gal (45.5l) fuel tank; 5.50x15 (front), 6.00x15 (rear) tyres; centre-lock ventilated disc wheels. PRODUCTION 1.

1958-1960 MGA Twin Cam

A double overhead camshaft conversion of the already obsolescent B-series engine was the inspiration of talented Jowett Javelin and returnee Nuffield designer Gerald Marley Palmer (1911-1999). The classic racing layout, which dated back to 1911 the work of Ernest Henry and colleagues Zuccarelli, Boillot and Goux at Peugeot with twin overhead camshafts was still rare for production engines on account of cost, noise and complication. Jaguar managed it with chains in an era before toothed belt drives, and MG's chain drive off a jackshaft, located where the B-series camshaft was on the regular engine, proved bold. The valves were inclined at 80deg, the crankshaft stiffened and there were stronger con-rods. Polished alloy valve covers made a brave sight but accessibility was problematical. By the time it was put into production Palmer, who had been largely responsible for the ZA and ZB

Magnette, had fallen out with LP Lord and left for Vauxhall, never understanding why the endemic piston problems that dogged it were not put right by Morris Engines. Development was left to Twin Cam owners who learned, eventually, how to maintain reliable oil-tight units. Dearth of 100 octane fuel was a problem, and the Twin Cam had to be withdrawn owing to heavy warranty claims. The engine was noisy and oil consumption of a pint every 120miles (193km) not unusual. The Twin Cam was in trouble from the start, four cars on the press launch at Chobham overheated, there were problems with running-on and one was crashed. Car journalists never quite forgave the Twin Cam for what they saw as indifferent quality and unreliability, yet it remained at a premium in the classic market.

BODY sports, 2 doors, 2 seats; weight 2184 lb (991kg). ENGINE 4 cylinders, in-line; front; 75.4mm x 89mm, 1588cc; compr 9.9:1 (later reduced to 8.3); 108bhp (80.5kW) @ 6700rpm; 68bhp (50.7kW)/l; 104lbft (140Nm) @ 4500rpm. ENGINE STRUCTURE two chain-driven overhead camshafts; cast iron block, aluminium cylinder head; finned aluminium sump 2 SU H6 carburettors, 12-volt coil ignition; SU electric fuel pump; 3-bearing crankshaft; engine rubber-mounted. TRANSMISSION rear wheel drive; Borg & Beck sdp 8in (20.3cm) clutch; 4-speed manual gearbox, synchromesh 2nd, 3rd, top; Hardy Spicer prop shaft; hypoid bevel final drive 4.3:1. CHASSIS steel perimeter frame with scuttle reinforcement; ifs by coil spring and unequal length wishbone; live rear axle with semi-elliptic springs; Armstrong lever arm dampers front and rear; Dunlop disc brakes; Cam Gears rack and pinion; 10gal (45.5l) fuel tank; 5.60–15; centre-lock ventilated disc wheels. DIMENSIONS wheelbase 94in (239cm); track 47.5in (121cm) front, 48.8in (20.3cm) (124cm) rear; turning circle 28ft (8.5m); ground clearance 6in (15cm); length 156in (396cm); width 58in (20.3cm) (147cm); height 50in (127cm). PERFORMANCE maximum speed 115mph (185kph); 17mph (27.4kph) @ 1000rpm; 0-60mph (96kph) 9.1sec; 9.2kg/bhp (12.3kg/kW); fuel consumption 22.2mpg (12.7L/100km). PRICE £1195 7s 6d (£1195.37.5p), coupe £1281 15s 10d (£1281.79p). PRODUCTION 2111.

Below: Dunlop disc brakes and centre-lock perforated steel wheels completed the Twin Cam's specification, going some way in justifying its higher price. British Motor Museum Gaydon cutaway exhibit.

1959-1961 Magnette Mark III

Scant interest was shown by MG devotees in the Farina-styled range of 1½ litre BMC saloons. Boxy and bland, they marked the nadir of British Motor Corporation design. *Motor Sport* was unequivocal: "The Mark III Magnette does not fill the individualistic niche, which caused enthusiasts to regard the ZB Magnette with such warm affection. It remains (to be seen) whether one might as well just order a 15/60 Wolseley or an Austin A55." It could have added a Morris Oxford, or a Riley 4/68; they were all the same. The best *Motor Sport* could find to say about the Magnette was on the torsion bar boot lid support that dispensed with a prop, and the way the direction flasher warning lights dimmed after dark. *The Motor* was critical of 17 points requiring lubrication by grease-gun every 1,000 miles: "an anachronism on a modern design." Too true. Testers were not impressed with oil consumption of a pint every 200 miles. Taking the excitement out of a car in which it was traditionally such a vital ingredient was ill-conceived; the regression to cam and lever from rack and pinion steering was particularly inept. "The steering is better at high speed than low ones, when apparent friction at some point in the linkage makes it too heavy for parking and causes slight weaving at town speeds." The Magnette's price of £200 over the Morris and Austin equivalents gave buyers a livelier turn of speed and superior furnishings but not much else. Even the passing of years did not offer much succour for such a heavy application of the derided principles of badge engineering.

BODY saloon, 4 doors; 4 seats; weight 2407.9lb (1092.2kg). ENGINE 4 cylinders, in line; front; 73.025mm x 88.9mm, 1489cc; compr 8.3:1; 66.5bhp (49.6kW) @ 5200rpm; 44.7bhp (33.3kW)/l; 85lbft (114Nm) @ 3300rpm. ENGINE STRUCTURE pushrod overhead valve, chain-driven camshaft; cast iron cylinder block and head; 2 SU HD 4 carburettors; SU PD electric fuel pump; centrifugal and vacuum ignition; 3-bearing crankshaft. TRANSMISSION rear wheel drive; Borg & Beck 8in (20.3cm) sdp clutch; 4-speed gearbox, synchromesh; Hardy Spicer open prop shaft; hypoid bevel final drive 4.3:1. CHASSIS DETAILS steel monocoque structure; ifs by coil springs and wishbones; live axle and semi-elliptic springs; Armstrong lever arm hydraulic dampers; Girling hydraulic 9in (22.9cm) (22.9cm) drum brakes (2LS front); Cam and lever steering; 10.5gal (47.7l) fuel tank; 5.90–14 tubeless tyres; steel disc wheels. DIMENSIONS wheelbase 99.25in (252cm); track 50in (127cm) front, 48.5in (123cm) rear; turning circle 34.5ft (10.5m); grnd clrnce 7in (17.8cm) (17.8cm); length 178in (20.3cm) (452cm); width 63.25in (160.7cm); height 59.25in (150.5cm). PERFORMANCE maximum speed 87.7mph (140.8kph); 16.6m (26.7kph) @ 1000rpm; 0-60mph (96kph) 19.7sec; 16.4kg/bhp (22kg/kW); fuel consumption 26.8mpg (10.5l/100km). PRICE £1012 12s 6d (£1012. 62.5p). PRODUCTION 15,676.

1959-1961 MGA 1600 Tourer and Coupe

After earning $60 million from exports to the United States, increasing the MGA's B-series engine's capacity to match the Twin Cam's 1588cc was no more than a simplification of cylinder block machining. It brought an extra 6bhp (4.5kW) or some 10%, and a 17% increase in torque. The engine was no quieter and fuel consumption suffered slightly but acceleration was swifter and even the open version, production of which began in May, could now reach a genuine 100mph. Lockheed disc brakes were fitted to the front wheels, but without a servo they were on the heavy side.

It was all a bit late to recapture the Triumph and Austin-Healey market for top speed although MG still had the edge in handling and pedigree. Among improvements expected to deal with competitors offering better weather protection, the tourers now had sliding sidescreens of clear plastic laminate instead of soft transparent material. The old ones flopped about, soon scratched and became opaque. Plastic detachable hardtops proliferated; the works one was well proportioned and fitted snugly. Luggage space remained at a premium however, and on 1600 Coupes the spare wheel in the boot was moved rearwards, providing more suitcase room within the carpeted area behind the seats. Coupes were made complete and then the top was welded on at Morris Bodies in Cowley. They had better furnishings and a fabric-covered facia.

BODY sports, 2 doors, 2 seats; weight 1988lb (902kg), Coupe 2105lb (955kg). ENGINE 4 cylinders, in-line; front; 75.39mm x 88.9mm, 1588cc; compr 8.3:1; 79.5bhp (59.3kW) @ 5600rpm; 50.1bhp (37.3kW)/l; 87.4lbft (117Nm) @ 3800rpm. ENGINE STRUCTURE pushrod overhead valve; chain-driven camshaft; cast iron cylinder head and block; 2 SU inclined carburettors, 12-volt coil ignition; SU electric fuel pump; 3-bearing crankshaft; engine rubber-mounted TRANSMISSION rear wheel drive; Borg & Beck sdp 8in (20.3cm) clutch; 4-speed manual gearbox, synchromesh 2nd, 3rd, top; hypoid bevel final drive 4.3:1; 4.55 optional. CHASSIS DETAILS steel perimeter frame with scuttle reinforcement; ifs by coil spring and unequal length wishbone; live rear axle with semi-elliptic springs; Armstrong lever arm dampers front and rear; Lockheed hydraulic brakes 11in (27.9cm) (27.9cm) discs at front, 10in (25.4cm) (25.4cm) drums at back, 2LS at front; rack and pinion steering; 10gal (45.5l) fuel tank; 5.60–15; centre-lock wire wheels optional. DIMENSIONS wheelbase 94in (238.8cm); track 47.5in (120.7cm) front, 48.8in (20.3cm) (123.8cm) rear; turning circle 28ft (8.5m); ground clearance 6in (15.2cm); length156in (396.2cm); width 58in (20.3cm) (147.3cm); height 50in (127cm). PERFORMANCE maximum speed 100.5mph (161.7kph); 17mph (27.4kph) @ 1000rpm; 0-60mph (96kph) 15.0sec; 11.3kg/bhp (15.2kg/kW) tourer, 12kg/bhp (16.1kg/kW) coupe; fuel con 27mpg (10.5l/100km). PRICE £940 7s 6d (£940 37.5p), Coupe £1026 15s 10d (£1026.79). PRODUCTION 31,501.

1960 Ted Lund-Colin Escott Le Mans coupe

Thoroughly alarmed by the heavy rate of racing accidents that seemed to pursue the works MGAs, the BMC directors withdrew official backing. In future the factory would take part only in rallies, which seemed safer, and while the new Competitions Department based at Abingdon would emerge with credit campaigning Minis and Austin-Healeys as well as MGs, racing had always been in MG's heritage. One subterfuge to get back to Le Mans was the factory building one car with a twin-cam engine as tried out in the 1955 TT. It was road-registered SRX 210 and passed to Ted Lund, a former member of the factory Le Mans team.

The body surreptitiously incorporated some of the aluminium body components made for the 1955 Le Mans cars, but it was invariably described as a private venture and was entered for the Le Mans Grand Prix d'Endurance in 1959, 1960, and 1961 by the North West Centre of the MG Car Club. At speed in 1959 Co-driver Colin Escott hit a dog at night, putting paid to the car's cooling system (and the dog). For 1960 Lund formally bought the car from the factory, it was equipped with Coupe doors and fastback bodywork, won the 2.0 litre class and finished 12th at 91.2mph (146.4kph); a creditable performance. The versatile vehicle came back in 1961 more highly stressed, and lapping at 101.6mph (163.1kph) proved too much for it; the engine failed early in the race. It then went on to a long career in club racing, a distinctive survivor of MG's official support for motor racing.

BODY sports, 2 seats; weight of 1959 open car approx 1600lb (726kg). ENGINE 4 cylinders, in-line; front; 75.4mm x 89mm, 1588cc, for 1960 79.4mm x 89mm, 1763cc; for 1961 128bhp (95.5kW); 72.6bhp (54.1kW)/l. ENGINE STRUCTURE two chain-driven overhead camshafts; cast iron block, aluminium cylinder head; finned aluminium sump; two twin-choke Weber carburettors; 12-volt coil ignition; two SU electric fuel pumps; 3-bearing crankshaft. TRANSMISSION rear wheel drive; Borg & Beck sdp 8in (20.3cm) clutch; 4-speed manual close-ratio gearbox, synchromesh 2nd, 3rd, top; hypoid final drive with bevel-type diff 4.1:1. CHASSIS DETAILS steel perimeter frame with scuttle reinforcement; ifs by coil spring and unequal length wishbone; live rear axle with semi-elliptic springs; Armstrong lever arm dampers front and rear; Lockheed hydraulic brakes 10in (25.4cm) drums, 2LS at front; rack and pinion steering; 20gal (90.91) fuel tank; front 5.60–15tyres, rear 6.00-15; centre-lock steel wheels. DIMENSIONS wheelbase 94in (238.8cm); track 47.5in (120.7cm) front, 48.75in (123.8cm) rear; turning circle 28ft (8.5m); ground clearance 6in (15.24cm); length 150in (381cm); width 58in (20.3cm) (147.3cm); height 41in (104.1cm). PERFORMANCE 1959, approximately 125mph (201kph) with metal cover on passenger seat and racing screen; 1960, 130mph (209kph); 1961, 140mph (225kph). PRODUCTION 1.

1960s EX220 and ADO34 front wheel drive MG prototype

The advent of the Mini changed perceptions of sporting handling and road holding. Aspiring racing and rally enthusiasts turned to Minis and Mini-Coopers because they could often outperform contemporary sports cars of greater power. Front wheel drive could out-corner and outmanoeuvre old-style rear-drive in round-the-pylons driving test competitions popular with car clubs. Even so, an MG sports car with front wheel drive and all the attributes of the new wave of clubmens' ideals was too radical to reach production. Drawings were started even before the Mini was launched and plans laid by Syd Enever in 1959 looked like a shrunken version of the not-yet-announced MGB. It was seriously considered, several experimental schemes were drawn up and modeled. EX220 was subsequently awarded an Austin (or Amalgamated) Drawing Office (ADO) code for development and since it survived was among the best known prototypes. A sports car using the Mini's transverse engine would have suited MG's tradition of corresponding to volume-produced cars and there was also a Coupe prototype. Alec Issigonis believed a sports car should have rear wheel drive, however, and opposed it. Technical difficulties with scuttle shake owing to the absence of a roof could have been overcome.

One running prototype was based on Minivan underpinnings with a deep central backbone tunnel but in the end this too survived only as a museum footnote.

BODY roadster, 2 doors, 2 seats. ENGINE 4 cylinders; front, transverse; 64.58mm x 83.72mm, 1098cc; compr 8.9:1, 8.0 optional; 56bhp (41.7kW) @ 5500rpm; 51bhp (38kW)/l; 62lbft (83Nm) @ 3250rpm. ENGINE STRUCTURE pushrod overhead valve, chain-driven camshaft; cast iron cylinder head, block; 2 SU HS2 1¼ in carburettors, SU electric fuel pump; 3-bearing crankshaft. TRANSMISSION front wheel drive; sdp clutch; 4-speed manual gearbox, synchromesh; helical gear final drive 4.22:1. CHASSIS DETAILS independent suspension by Hydrolastic cone rubber springs; telescopic dampers; hydraulic disc and drum brakes; rack and pinion steering. PRODUCTION 2.

Below: Pininfarina designed a stylish body, but the sporty Mini was destined to be the Mini-Cooper, not an open 2-seater. Gaydon British Motor Museum exhibit.

1961-1962 MGA 1600 Mk II Tourer and Coupe

The next change in MGA engine size was carried out, like the previous one, partly to achieve economies in machining production time. In April the 1600 was replaced by a 1600 Mk II, identifiable by vertical rather than sloping grille bars. The increase from 1588cc to 1622cc meant new cylinder block and head castings and, although the 88.9mm stroke never altered, the new 76.2mm bore meant two pairs of cylinders were conjoined, with porting and airflow improved. The combustion chambers were reshaped with bigger inlet valves, and the concave pistons became flat-topped. Comfortingly for old-style engineers the cylinder dimensions were now memorably 3in x 3½in. Time was however catching up with the MGA. Softly sprung cars like the Sunbeam Alpine were appearing and while their handling and roadholding were not in the same class, their ride and roadworthiness had to be somehow matched. "By modern standards," said *The Autocar* in October 1959, "the MGA's suspension must be considered firm; on smooth roads this is no disadvantage ... this does not apply on rougher surfaces however, and a feeling that the wheels were hopping and jumping, accompanied by intermittent tyre squeal indicated that the tyres were not maintaining full contact with the road." It had taken the best part of a decade for sporting drivers to compromise over firm springing and an even ride; the competition had just about shown it was possible to have both. Out of 101,081 made between 1955 and 1962, 91,547, or 91 per cent of MGAs, had been exported. America took 81,153, a successful score.

The MGA had an encouraging debut in competition, when a Coupe driven by Don and Erle Morley won its class and finished 28th in the 1962 Monte Carlo Rally.

BODY sports, 2 doors, 2 seats; weight 1985lb (900kg); coupe 2045lb (928kg). ENGINE 4 cylinders, in-line; front; 76.2mm x 88.9mm, 1622cc; compr 8.9:1; 90bhp (67.1kW) @ 5500rpm; 55.5bhp (41.4kW)/l; 97.4lbft (131Nm) @ 4000rpm. ENGINE STRUCTURE pushrod overhead valve; chain-driven camshaft; cast iron cylinder head and block; 2 SU inclined carburettors, 12-volt coil ignition; SU electric fuel pump; 3-bearing crankshaft; engine rubber-mounted TRANSMISSION rear wheel drive; Borg & Beck sdp 8in (20.3cm) clutch; 4-speed manual gearbox, synchromesh 2nd, 3rd, top; hypoid bevel final drive 4.1:1. CHASSIS DETAILS steel perimeter frame with scuttle reinforcement; ifs by coil spring and unequal length wishbone; live rear axle with semi-elliptic springs; Armstrong lever arm dampers front and rear; Lockheed hydraulic brakes 11in (27.9cm) discs at front 10in (25.4cm) drums rear; rack and pinion steering; 10gal (45.5l) fuel tank; 5.60–15; centre-lock wire wheels optional. DIMENSIONS wheelbase 94in (239cm); track 47.5in (121cm) front, 48.75in (124cm) rear; turning circle 28ft (8.5m); ground clearance 6in (15cm); length 156in (396cm); width 58in (20.3cm) (147cm); height 50in (127cm).PERFORMANCE maximum speed 101.4mph (162.8kph); 17.9mph (28.7kph) @ 1000rpm; 0-60mph (96kph) 13.7sec; 10kg/bhp (13.4kg/kW), coupe 10.3kg/bhp (13.8kg/kW); fuel consumption 22.3mpg (12.7l/100km). PRICE £940 7 6 (£940 37.5p). PRODUCTION 8,719.

1961-1962 MGA 1600 De Luxe; 1962 1600 De Luxe Mark II

Above: MGA owners often had external luggage racks to make up for the spare wheel taking up most of the niggardly boot.

Not for the first time when a new model failed to meet expectations, MG created something else to use up components that had been sanctioned and had to be paid for. After losing heart and withdrawing the Twin Cam, the 1600 De Luxe and Mark II De Luxe were introduced with the redundant body shells incorporating as many Twin Cam features as possible, except the troublesome engine. Two distinct Marks were made, 82 of them 1600 De Luxe models (70 Tourers and 12 Coupes) with 1588cc engines and 313 others 1600 De Luxe Mark IIs with 1622cc engines (290 Tourers and 23 Coupes). All were equipped with disc brakes, centre-lock ventilated pressed steel Dunlop wheels and leather facias, but there were wide differences in trim and specification as the parts bins emptied. These did not figure in official price-lists or catalogues, suggesting a damage limitation emergency that seems to have continued well into 1962 and the imminent introduction of the MGB. There was some uncertainty among production cars over what was to be included, some having optional equipment such as oil coolers, close ratio gearboxes and special seats fitted as standard. Many of them also seem not to have been sold through the usual channels, suggesting MG dealers and the sporting establishment were amongst eager clientele for some prized classics.

Spec 1600 De Luxe Mark II: BODY sports, 2 doors; 2 seats; weight 2184 lb (991kg). ENGINE 4 cylinders, in-line; front; 76.2mm x 88.9mm, 1622cc; cr 8.9:1; 90bhp (67.1kW) @ 5500rpm; 55.5bhp (41.4kW)/l; 97.4lbft (131Nm) @ 4000rpm. ENGINE STRUCTURE pushrod ohv; chain-driven camshaft; cast iron cylinder head and block; 2 SU inclined carburettors, 12-volt coil ignition; SU electric fuel pump; 3-bearing crankshaft; engine rubber-mounted. TRANSMISSION rear wheel drive; Borg & Beck sdp 8in (20.3cm) clutch; 4-speed manual gearbox, synchromesh 2nd, 3rd, top; Hardy Spicer prop shaft; hypoid bevel final drive 4.3:1. CHASSIS DETAILS steel perimeter frame with scuttle reinforcement; ifs by coil spring and unequal length wishbone; live rear axle with semi-elliptic springs; Armstrong lever arm dampers front and rear; Dunlop disc brakes; Cam Gears rack and pinion steering; 10gal (45.5l) fuel tank; 5.60–15; centre-lock ventilated disc wheels. DIMENSIONS wheelbase 94in (239cm); track 47.5in (121cm) front, 48.75in (124cm) rear; turning circle 28ft (8.5m); ground clearance 6in (15cm); length156in (396cm); width 58in (20.3cm) (147cm); height 50in (127cm). PERFORMANCE maximum speed 102mph (163.7kph); 17mph (27.4kph) @ 1000rpm; 0-60mph (96kph) 15.5sec; 11kg/bhp (14.8kg/kW); fuel consumption 22.2mpg (12.7l/100km). PRICE see text. PRODUCTION 395.

Above: Generous access to B-series engine.

Below: Recognition feature, recessed grille with upright slats

1961-1968 Magnette Mk IV

Widespread approval for the Farina Magnette took a long time. It had failed to produce the affection and respect felt for the ZA and ZB let alone the Farina Mk III, yet in its latest form with some useful detail and cosmetic improvements it became almost worthy. The engine was increased in size and power by rounding off the bore to 3in (76.2mm). The stroke was 3.5in (88.9mm) and the result was a steeper torque curve and an all-round improvement in speed and liveliness with little effect on fuel consumption. The fuel tank was still too small to provide a useful range; something like 250 miles was about its limit. Doubling the valve springs diminished but did not altogether cure valve bounce at high engine speeds.

One major innovation - the Mark IV Magnette was the first MG to be offered with fully automatic transmission. Ride and handling were improved by the addition of anti-roll bars at front and back, as well as an increase in wheelbase and track. *The Autocar* was pleased at having already offered solutions to the car's "...somewhat bucking ride." After testing the Mark III it changed its view: "Today's Magnette is well up to standard for a conventional design with a rigid back axle, offering its rear-seat occupants almost as easy travelling as those in the front." What it called "lost movement" in the steering was however a persistent shortcoming.

BODY saloon; 4 doors; 4 seats; weight 2534lb (1150kg). ENGINE 4-cylinders; in-line; front; 76.2mm x 88.9mm, 1622cc; compr 8.3:1; 68bhp (50.7kW)@ 5000rpm; 41.9bhp (31.3kW)/l; 89lb ft (119Nm) @ 2500rpm. ENGINE STRUCTURE pushrod overhead valve, chain-driven camshaft; cast iron cylinder block and head; 2 SU H4 carburettors; SU electric fuel pump; centrifugal and vacuum ignition; 3-bearing crankshaft. TRANSMISSION rear wheel drive; Borg & Beck 8in (20.3cm) sdp clutch; 4-speed manual gearbox, synchromesh; optional Borg Warner Type 35 automatic; Hardy Spicer open prop shaft; hypoid bevel final drive 4.3:1. CHASSIS DETAILS steel monocoque structure; ifs by coil springs and wishbones; live axle and semi-elliptic springs; anti-roll bars front and rear; Armstrong lever arm hydraulic dampers; Girling hydraulic 9in (22.9cm) drum brakes (2LS front); Cam and lever steering; 10.5gal (47.7l) fuel tank; 5.90– 14tubeless tyres; steel disc wheels. DIMENSIONS wheelbase 100.2in (254.5cm); track 50.2in (127.5cm) front, 51.4in (130.5cm) rear; turning circle 36ft (11m); ground clearance 7in (17.8cm) (17.8cm); length 178in (20.3cm) (452cm); width 63.25in (160.7cm); height 59.25in (150.5cm). PERFORMANCE maximum speed 86mph (138kph); 16.6mph (26.7kph) @ 1000rpm; 0-60mph (96kph) 19sec; 16.9kg/bhp (22.7kg/kW); fuel consumption 25.2mpg (11.2l/100km). PRICE £997. 17s 9d (£997 88.75p). PRODUCTION 13,738.

1961-1962 Midget GAN1

Above: original BMC publicity picture has at top left the ancestor, a Mark I Austin-Healey Sprite with winter hardtop.

Any lingering feelings over LP Lord's adoption of the Austin-Healey 100 in what had looked like Abingdon's hour of need in 1952 were assuaged when the Abingdon factory undertook the assembly of Austin-Healeys. There was little opportunity for rancour when the Sprite was adopted as the basis for a new MG Midget. Purloining a name from a Riley of 1936-1938 that was still on Nuffield-BMC's catalogue of titles, the Sprite was in the idiom of the original series of Midgets, borrowing components from a small production saloon, the Austin A35 and splicing them into a plain, simple and easily manufactured open 2-seater. Donald Healey was in the Kimber mould of practicality, salesmanship and achievement.

The result was probably the car MG ought to have been building all along and soon it had the cheerful frog-eyed 2-seater with its precise steering and nimble handling on the production line at Abingdon. In 1961 as sales of the Mark I Sprite began to falter BMC decided the shortage of luggage room and pert visage had no place in the modern world and ordered a re-skin, also seizing the opportunity to co-produce it as an MG. In the Byzantine world of BMC the Healeys at Warwick were tasked with producing a new front for it and Syd Enever at Abingdon a new rear. It was a wonder it ever got made at all, far less matching both ends to produce the Mark II Sprite coded HAN6 and Mark I Midget GAN 1. The conversion was scarcely lavish. The curve of the old frog-eye bonnet line to the scuttle remained, but on the whole it was effective and provided the design with a lifespan the ingenious original Sprite had been unlikely to enjoy on its own.

BODY sports, 2 doors, 2 seats; weight 1566lb (710kg). ENGINE 4 cylinders, in-line; front; 62.94mm x 76.2mm, 948cc; cr 9.0:1; 46bhp (34.3kW) @ 5500rpm; 48.5bhp (36.2kW)/l; 52.8lbft (71Nm) @ 3000rpm. ENGINE STRUCTURE pushrod overhead valve, chain-driven camshaft; cast iron cyl head, block; 2 SU HS2 1¼ in carbs, mechanical fuel pump; 3-bearing crankshaft. TRANSMISSION rear wheel drive; 6¼ in sdp clutch; 4-speed manual gearbox, synchromesh; hypoid final drive 4.22:1. CHASSIS DETAILS steel platform chassis; ifs by coil springs and wishbones; live rear axle quarter-elliptic 15-leaf springs, radius arms, lever arm hydraulic dampers; Lockheed hydraulic brakes 7in (17.8cm) drums; rack and pinion; 6gal (27.3l) tank; 5.20–13 Dunlop tubeless tyres; ventilated steel disc wheels, wire wheels optional. DIMENSIONS wheelbase 80in (203.2cm); track 45.25in (114.9cm) front, 44.75in (113.7cm) rear; turning circle 30.1ft (9.2m); ground clearance 5in (12.7cm); length135.25in (346.1cm); width 53in (134.5cm); height 49.75in (126.4cm). PERFORMANCE max speed 84.7mph (136kph); 15.4mph (24.7kph) @ 1000rpm; 0-60mph (96kph) 20.2sec; 15.4kg/bhp (20.7kg/kW); fuel consumption 33.4mpg (8.5l/100km). PRICE £669 15s 10d (£669 79.16p). PRODUCTION 16,084.

1962-1964 MG Midget GAN 2

Even though Purchase Tax was increased in 1961, the Midget introduced in June remained under £700, about the price of an Austin A40, Ford Anglia or Morris Minor 4-door. The only real competition came from the Austin-Healey Sprite at £660 or the Triumph Herald Coupe at £729 or Convertible at £776. The unlovely Turner A35 Sports was £839. There was nothing remotely to match the open 2-seater Midget/Sprite whose differences were purely cosmetic. The MG had fancy wheel trim, a prettier grille, chrome strips on the flanks, carpet on the floor instead of ribbed rubber and the cachet of being an

MG, although even that was open to argument. Austin-Healey enthusiasts could be fiercely loyal and looked upon the Midget as essentially their car, titivated as an MG and not engineered at Abingdon. The first revision came at the Earls Court London Motor Show in October 1962 with a sort of Mark 1½ when the capacity of the BMC A-series engine was increased, putting the performance up to the level of the Riley 1.5 or Alfa Romeo Giulietta ti. It was still not fast in absolute terms but the new engine, disc brakes, and gearbox with baulk-ring synchromesh gave it a boost at a time when the traditional sports car was being challenged by small front wheel drive saloons like the Mini-Cooper. A padded roll was added to the facia coaming as a token safety feature.

BODY sports, 2 doors, 2 seats; weight 1566lb (710kg). ENGINE 4 cylinders, in-line; front; 64.58mm x 83.72mm, 1098cc; compr 8.9:1, 8.0 optional; 56bhp (41.7kW) @ 5500rpm; 51bhp (38kW)/l; 62lb ft (83Nm) @ 3250rpm. ENGINE STRUCTURE pushrod ohv, chain-driven camshaft; cast iron cylinder head, block; 2 SU HS2 1¼ in carburettors, SU electric fuel pump; 3-bearing crankshaft. TRANSMISSION rear wheel drive; 7¼ in sdp clutch; 4-speed manual gearbox, synchromesh; hypoid final drive 4.22:1. CHASSIS DETAILS steel platform chassis; ifs by coil springs and wishbones; live rear axle quarter-elliptic 15-leaf springs, radius arms, lever arm hydraulic dampers; Lockheed hydraulic brakes 8¼ in (21cm) discs front, 7in (17.8cm) drums rear; rack and pinion steering; 6gal (27.3l) fuel tank; 5.20–13 Dunlop tubeless tyres; ventilated steel disc wheels; wire wheels optional. DIMENSIONS wheelbase 80in (203.2cm); track 45.25in (114.9cm) front, 44.75in (113.7cm) rear; turning circle 30.1ft (9.2m); ground clearance 5in (12.7cm); length135.25in (346.1cm); width 53in (134.5cm); height 49.75in (126.4cm). PERFORMANCE maximum speed 88.6mph (142.2kph); 15.4mph (24.7kph) @ 1000rpm; 0-60mph (96kph) 16.9sec; 12.7kg/bhp (17kg/kW); fuel consumption 29.9mpg (9.4l/100km). PRICE 1962 £598 13s 9d (£598 69p). PRODUCTION 9601.

Left: Ubiquitous A-series with lifting lugs and heat shield behind the carburettors used almost throughout the life-span of the Midget.

1962-1968 MG1100

Above: Two-colour paintwork was an option and there was a two-door version that went for export.

The 1100 was not the first MG with Issigonis ingredients. The front suspension had also come off his drawing board in the 1930s so it was something of a veteran against the new front wheel drive transverse-engined Hydrolastic-suspended saloons that became one of the mainstays of BMC. It was a series not, in the end, an unalloyed successes, but the prescription produced three million of them and for many years they were among the top-sellers in Britain. The principle of MGs being twin-carburettored, leather upholstered versions of others in a similar series became well established. It was a logical progression from the 1930s Nuffield Group when big MG saloons borrowed components from Wolseley and Morris but it somehow became more obvious in the 1960s. Chrome stripes and dummy radiator grilles fooled nobody. Customers knew perfectly well the cars were coming off production lines a long way from Abingdon. William Morris's old principle of assembling cars from components "made by experts" elsewhere held good only so long as a host plant applied the finish and sustained a tradition. Rolls-Royces were no less Rolls-Royces so long as they came from the factory in Crewe even though the bodies were made by Pressed Steel at Cowley. They would not have been Rolls-Royces had they been wholly made at Cowley and in the long run much the same went for MGs.

BODY saloon, 2/4 doors, 4 seats; weight 1820lb (825.5kg). ENGINE 4 cylinders; front, transverse; 64.58mm x 83.72mm, 1098cc; compr 8.9:1, optional 8.1:1; 55bhp (41kW) @ 5500rpm; 50.1bhp (37.3kW)/l; 61lbft (82Nm) @ 2750rpm. ENGINE STRUCTURE pushrod overhead valve, chain-driven camshaft; cast iron cylinder head, block; 2 SU HS2 1¼ in carburettors, SU electric fuel pump; centrifugal and vacuum ignition; 3-bearing crank.

TRANSMISSION front wheel drive; 7.125in sdp clutch; 4-speed manual gearbox, synchromesh; helical gear final drive 4.133:1. CHASSIS steel monocoque structure; independent suspension by interconnected Hydrolastic units and rubber-cone springs; transverse wishbone at front, trailing arms at rear with anti-toll bar and two torsion bars for pitch stiffness; Lockheed brakes, 8in (20.3cm) disc front 8in (20.3cm) drum rear with pressure limiting valve; Cam Gears rack and pinion; 8gal (38.4l) fuel tank; 5.50-12 tubeless tyres. DIMENSIONS wheelbase 93.5in (237.5cm); track 51.5in (130.8cm) front, 50.75in (128.9cm) rear; turning circle 32.5ft (9.9m); grnd clearance 5.25in (13.3cm); length 146.75in (372.7cm); width 60.25in (153cm); height 53.25in (135.3cm). PERFORMANCE max speed 87.3mph (140kph); 14.9mph (23.9kph) @ 1000rpm; 0-60mph (96kph) 20.2sec; 15kg/bhp (20.1kg/kW); 30mpg (9.4l/100km). PRICE £713 9s 7d (£713 48p). PRODUCTION 116,827.

Below: Demo picture of the Hydrolastic suspension seemed convincing.

1962-1967 MGB

By the 1960s an MG that could keep up with Triumph TRs and Austin-Healeys was urgently needed and in the light of the Mini and front wheel drive transverse-engined leadership looked technically progressive. Experimental initiatives with MGA understructures were tried along with various engine configurations; Hayter made the bonnet aperture wide enough to accommodate a 60 degree V4 that was under development. There was space for a V6. The MGA's 1622cc engine was not really strong enough and the Austin 2 litre heavy, so Charlie Griffin at Longbridge was called on for the Austin 1800 with a 3-bearing crank. Five bearings would come later. Independent and de Dion rear suspensions were tried but in reality MG was not eligible for the big budgets necessary for anything radical.

The result at the time seemed humdrum and worthy rather than epoch-making yet it stood the test of time astonishingly well. It may not have quite kept up with the opposition in terms of speed and acceleration but was well up to the mark in comfort and accommodation. As a pioneer of open 2-seat monocoques, it was well ahead. The MGB may have been too robust for its power-to-weight ratio, but it solved the problems of a structure with wide door apertures, a serious threat to its beam strength, far better than almost any of its successors. Plenty of other manufacturers never came close. MGBs never suffered from the crippling scuttle shake that bedeviled others, rattling themselves to pieces in the next half century. It never had the crisp handling or nimbleness of the MGA, but it was strong, safe and stylish, adhering to the prescription laid down by Cecil Kimber half a century before. True to this tradition it also kept its links to volume production components endearing it to generations of private owners who had easy and cheap access to a virtually inexhaustible supply of volume-produced engines gearboxes and axles. The MGB had a well-tried engine and familiar, relatively cheap running gear. An entire industry was created in its afterlife, dedicated to its restoration and preservation.

Don Hayter in his MGB Story describes how the body profile evolved from EX 181 the record-breaking Roaring Raindrop of 1957. Hayter started with the classic long-tailed teardrop shape until, much as aerodynamic pioneer Viennese Paul Jaray founder of the Stromlinien Karosserie Gesellschaf had in the 1930s, modified it for practicalities such as air intakes and occupants. Also, like Stuttgart Professor Wunibald Kamm, Hayter cut short EX 181's long tail. The result was the exquisitely proportioned B that made it not only an aesthetic but also an engineering accomplishment.

A success from the start, probably a little over-engineered and over-weight the MGB underwent a mid-life crisis in the dog days of British Leyland. It proceeded downhill to its demise in 1980 with the afflictions of the elderly, lethargic perhaps and feeble constricted by emission and safety legislation poorly applied and under-invested by the indifferent and declining management that oversaw it.

BODY roadster, 2 doors, 2 seats; weight 2030lb (921kg). ENGINE 4 cylinders, in-line; front; 80.26mm x 88.9mm, 1798cc; compr 8.8:1; 95bhp (70.8kW) @ 5400rpm; 52.8bhp (39.4kW)/l; 110lbft (147.5Nm) @ 3000rpm. ENGINE STRUCTURE pushrod overhead valve; chain-driven camshaft; cast iron cylinder head and block; 2 SU inclined H4 carburettors, SU electrical fuel pump; centrifugal and vacuum ignition control; 3-bearing crankshaft; engine rubber-mounted; optional oil cooler. TRANSMISSION rear wheel drive; Borg & Beck 8in (20.3cm) diaphragm spring clutch; 4-speed manual, synchromesh; single-piece open prop shaft; hypoid bevel final drive 3.91:1. CHASSIS steel monocoque structure; ifs by coil springs and unequal wishbones; live axle with semi-elliptic springs; Armstrong lever arm dampers; Lockheed hydraulic brakes, front 10.75in (27.3cm) discs, rear 10in (25.4cm) drums; rack and helical pinion steering; 10gal (45.5l) fuel tank; Dunlop Road Speed 5.90-14 tyres; 4J rims; wire wheels optional. DIMENSIONS wheelbase 91in (231.1cm); track 49in (22.9cm) (124.5cm); turning circle 30.5ft (9.3m); ground clearance 4.25in (10.8cm); length 153.75in (390.5cm); width 60in (152.4cm); height 49.25in (125cm). PERFORMANCE maximum speed 108.1mph (173.5kph);16.4mph (26.3kph) @ 1000rpm; 0-60mph (96kph) 12.1sec; 9.7kg/bhp (13kg/kW); fuel consumption 23.0mpg (12.3l/100km). PRICE £870 2s 3d (£870 30p). PRODUCTION 513,276 all MGB.

1964-1966 Midget MkII GAN 3

Changes to what was really a third Midget/Sprite specification, but known as the Midget Mark II, were more fundamental than with the Mark 1 and 1½ GAN2 of 1962. Sports cars of the 1960s still handled much as they had in the 1950s with sharp steering, a firm ride resulting from small spring amplitudes and little body roll. Essentially rear wheel drive over-steerers, they were being caught out by small front wheel drive cars which, although not all smooth-riding, captured much of the clubman sporting market. The Spridget had to change from quarter-elliptic rear springs, which had used the bump stops as an integral feature of the suspension, to semi-elliptics following the addition of the rear luggage space. The change was not effected when the extended tail was introduced but only with the bigger engine and replaced the 14-leaf quarter-elliptic with a 5-leaf half-elliptic. The radius arm of the old arrangement was no longer deemed necessary so although the ride improved the axle's longitudinal location was less strict. There was a small reduction in unsprung weight through this and the flatter ride was welcome in view of competition from the independently sprung but somewhat wayward Triumph Spitfire. The wider main crankshaft bearings gave better reliability and refinement, but led to a demand for improved furnishings such as the black crackle-finish facia with angled dials never, alas regarded with much affection. The plain original was durable and perfectly adequate.

As time went on seats changed all too often and a new windscreen and winding windows were another response to opposition from the Spitfire. The unfortunate corollary of the more secure and weatherproof windows was loss of elbow room. It resulted in a cockpit environment a far cry from the cowled scuttle and cutaway door of the TC and although a good deal drier and less draughty was charmless, bland and subject to the whims and fancies of marketing departments. Deckchair patterned upholstery and chrome geegaws came and, thankfully for the most part, duly went.

BODY sports, 2 doors, 2 seats; weight 1540lb (698.5kg). ENGINE 4 cylinders, in-line; front; 64.58mm x 83.72mm, 1098cc; compr 8.9:1, 8.0 optional; 59bhp (44kW) @ 5750rpm; 53.7bhp (40.1kW)/l); 62lbft (83Nm) @ 3250rpm. ENGINE STRUCTURE pushrod overhead valve, chain-driven camshaft; cast iron cylinder head, block; 2 SU HS2 1¼ in carburettors, SU electric fuel pump; 3-bearing crankshaft. TRANSMISSION rear wheel drive; 7¼ in sdp clutch; 4-speed manual gearbox, synchromesh; hypoid final drive 4.22:1. CHASSIS DETAILS steel platform chassis; ifs by coil springs and wishbones; live rear axle semi-elliptic leaf springs, Armstrong lever arm hydraulic dampers; Lockheed hydraulic brakes 8¼in (20.6cm) disc front, 7in (17.8cm) drums rear; rack and pinion steering; 6gal (27.3l) fuel tank; 5.20–13 Dunlop tubeless tyres; steel disc wheels; wire wheels optional. DIMENSIONS wheelbase 80in (203.2cm); track 45.25in (114.9cm) front, 44.75in (113.7cm) rear; turning circle 30.1ft (9.2m); ground clearance 5in (12.7cm); length 135.25in (346.1cm); width 53in (134.5cm); height 49.75in (126.4cm). PERFORMANCE maximum speed 91.8mph (147.4kph);15.4mph (24.7kph) @ 1000rpm; 0-60mph (96kph) 14.9sec; 11.8kg/bhp (15.8kg/kW); fuel consumption 29.2mpg (9.7;/100km). PRICE £622 17s 1d (£822 85.4p). PRODUCTION 26,601.

1964 EX234 Hydrolastic MG

Above and next page: Parade cars in the 1980 British Grand Prix included an EX234 prototype for Riccardo Patrese. They were finished running cars complete with interior trim and radio.

Abingdon never much liked the first Sprite's abrasive ride and even following its change to half-elliptics there were experiments aimed at improvements with BMC's Hydrolastic coupled fluid suspension. These began late in 1962 and carried on for some 18 months under EX229 and 231. Cost considerations led to the abandonment of a coil sprung independent back axle for the MGB but when proposals for its replacement were drawn up in 1964, Hydrolastic was once again considered. Another experimental model, EX234 could have, with different A-series and B-series engines, been capable of replacing both Midget and MGB and Pininfarina was commissioned for a stylish closed body with hints of MGB. A 2+2 prototype with an A-series engine and well-furnished interior including a fashionable central console was made, test-driven at Silverstone and produced encouraging results. Expert test drivers including John Surtees commended the roadholding. Roy Brocklehurst, who succeeded Syd Enever as chief engineer in 1971, worked in strictest secrecy on its underpinnings in the Abingdon boiler house. The prototype had an A-series engine driving the rear wheels through an Austin Gipsy final drive. Extensive experiments were carried out in an effort to adapt Hydrolastic suspension, used throughout the BMC front wheel drive range, to suit rear wheel drive but in the end the project was abandoned.

BODY roadster, 2 doors, 2 seats. ENGINE 4 cylinders; front, transverse; 70.6mm x 81.3mm, 1275cc; compr 9.75:1; 70bhp (52.2kW) @ 6000rpm; 54.9bhp (40.9kW)/l; 77lbft (103Nm) @ 3000rpm. Option for fitting 1798cc B-series. ENGINE STRUCTURE pushrod overhead valve, chain-driven camshaft; cast iron cylinder head, block; 2 SU HS2 1¼ in carburettors, SU electric fuel pump; centrifugal and vacuum ignition; 3-bearing crankshaft. TRANSMISSION front wheel drive; 7.125in sdp clutch; 4-speed manual gearbox, synchromesh; helical gear final drive 4.133:1. CHASSIS DETAILS steel monocoque structure; independent suspension by interconnected Hydrolastic units and rubber-cone springs; transverse wishbone at front, trailing arms at rear with anti-toll bar; sub-frame mounted; Lockheed brakes, 8in (20.3cm) disc front, 8in (20.3cm) drum rear with pressure limiting valve; Cam Gears rack and pinion steering; 8.5gal (38.6l) fuel tank; 13in wheels. DIMENSIONS wheelbase 80in (203.2cm). PRODUCTION 1.

1964 O-series MGB

For an engine whose origins went back to the dark days of the Second World War, the pushrod B-series had an astonishing lifespan. It was subject to continuous development along with a number of attempts at its replacement, such as the E-series built in a new plant at Cofton Hackett. This was a compact power unit developed for front wheel drive transverse engine cars and was designed to have two cylinders added for bigger cars. It would still have been difficult to fit into an MGB, however, and several alternatives were explored. A 2 litre version of the 5-bearing B-series designed as something of a stop-gap was the O-series, with an overhead camshaft aluminium cylinder head, which unfortunately took a long time to develop as US emission control measures grew more demanding. It was fitted in the Rover SD1 2000 in 1982 and tried experimentally in MGBs in the late 1970s. With the B approaching the end of its useful life however, it remained a tantalising might-have-been. The O-series evolved into the M16 twin overhead camshaft fuel injected Rover 820 engine, mooted as a possibility for the revived MGB of the 1990s. The V8 was chosen instead leaving a number of private owners, such as MG expert Roger Parker, to build one-offs that represented how the MGB might have been developed.

BODY roadster, 2 doors, 2 seats; weight approx 2442lb (1108kg) (GT). ENGINE 4 cylinders, in-line; front; 84.5mm x 89mm, 1996cc; compr 10.0:1; 118bhp (88.5kW) @ 5600rpm; 116bhp (86.44kW)/l; 120lbft (161Nm) @ 3500rpm. ENGINE STRUCTURE toothed belt driven twin overhead camshaft; aluminium cylinder head and block; 4-valves; fuel injection; contact-breaker ignition; 5-bearing crankshaft. TRANSMISSION rear wheel drive; sdp clutch; 5-speed manual synchromesh gearbox; hypoid bevel final drive. CHASSIS DETAILS steel monocoque structure; ifs by coil springs and unequal wishbones; live axle with semi-elliptic springs, anti-roll bars front and rear; Armstrong lever arm dampers; Lockheed hydraulic brakes with vacuum servo, front 10.75in (27.3cm) discs, rear 10in (25.4cm) drums; rack and pinion steering; 11gal (50l) fuel tank; 165SR -14 tyres; 5J rims. DIMENSIONS wheelbase 91in (231.1cm); track 49in (22.9cm) (124.5cm); turning circle 30.5ft (9.3m); ground clearance 5in (12.7cm); length 158.25in (402cm); width 60in (152.4cm); height 49.25in (125.1cm). PERFORMANCE 11kg/bhp (14.7kg/kW). PRODUCTION development cars only.

Above (both pictures): Author's project, an M-16 engined MGB Twin-Cam. Heritage bodyshell, Sherpa van bellhousing, and SD1 5-speed gearbox provided 120mph and lively acceleration.

The engine was turned from sideways in the 1990s front wheel drive Rover. A Sherpa van clutch housing bolted neatly into place, together with the 5-speed gearbox the Sherpa inherited from the Rover SD1. Drawing on pioneering work by Redditch MG specialist John Hill's MG Centre and Roger Parker's similar project, the tall overhead cam engine fitted the available space. The basis was a new body shell made by British Motor Heritage in its small factory at Faringdon, Oxfordshire on the original jigs and tools abandoned in 1980. While Rover had disdained the idea of a Twin-Cam MGB it took a keen interest in the project, helping instal the modern fuel injection and electronic engine management systems in place of the old-fashioned carburettors and ignition. The new wiring loom joined 1950s technology at one end to 1990s electronics at the other.

Moss Europe supplied components including wheels and suspension, and the MG Owners Club catalogue included the wood facia and door cappings which, had MG followed the example of Austin-Healey, would have given it the sort of interior customers expect of British classics. Rover senior designer Gordon Sked advised on the colour; Nordic Blue he suggested was a Rover colour that brought out the shape of the car. The 1960s chrome strip was done away with; the fixings were always a rust trap. A discreet coachline was more tasteful.

Tyres were important. When the MGB was designed most tyres were still layers of textiles. Michelin pioneered steel-bracing, which made the tread firmer and improved handling but in the 1960s suspension design had still to catch up. The Moss and Hopkinson modifications went some the way to make it ride and handle like a modern sports car, and Michelin MXT tyres deal with the enhanced performance - it did around 120mph.

When the MGB was announced *The Motor* road test, which the author helped compile in 1962, commented on a "marked bias towards a 'grand touring' character," and felt that at 18½ cwt MGBs were too heavy to win races. The extra power of the 2.0 litre engine compensated for this. The 'grand touring' character *The Motor* scoffed at was retained because great power was scarcely necessary in an open car.

1964 Dick Jacobs Midget Coupe

Following a long association with MG as a driver and, as proprietor of the Mill Garage at South Woodford, a dealer, Dick Jacobs was the means by which MG returned to the track in the 1960s. His racing Magnettes and MGA Twin Cams 1 and 2 MTW had been outstandingly successful, so when three Midgets were specially prepared for racing, two were assigned to him. The third went to Scottish enthusiast and MG works rally driver John Milne. Using a Midget platform as a basis, they had hand-made aluminium bodies secured with a mixture of rivets and epoxy-resin adhesive to create a stiff structure. They conformed roughly to the shape of a standard car with the addition of a sloping fastback roof and a rounded nose. Disc brakes, soon to be incorporated as standard, were also brought in and the suspension modified with high-rate springs and Armstrong competition shock absorbers, the back ones variable rate. Handling on the track was exemplary. Engines were enlarged as development continued with the BMC Formula Junior programme and driven by Alan Foster and Andrew Hedges the elegant little coupes gained a reputation for speed and reliability. They not only did well in British club racing, scoring a class win and third place in the 1963 Autosport Championship, but also in the 1964 100-kilometre race at Nürburgring, where they were first and second in class. Andrew Hedges and Roger Mac won their class in one at Sebring.

BODY coupe, 2 doors, 2 seats; weight 1242lb (563.4kg). ENGINE 4 cylinders, in-line; front; 63.73mm x 76.2mm, 972cc; compr 9.0:1; 75bhp (55.9kW) @ 5750rpm; 77.2bhp (57.5kW)/l. Various engines: 995cc, 1139cc, and 1293cc to comply with racing capacity classes. ENGINE STRUCTURE pushrod ohv, chain-driven camshaft; cast iron cylinder head, block; twin-choke Weber carb, mechanical fuel pump; 3-bearing crank, four-branch exhaust; machined polished combustion chambers and ports. Among engine variations was a 1293cc dry-sump unit for the Nürburgring. TRANSMISSION rwd; sdp clutch; 4-spd close-ratio gearbox, synchro; hypoid final drive 4.875:1. CHASSIS DETAILS steel platform chassis; ifs by coil springs and wishbones, anti-roll bar; live rear axle quarter-elliptic leaf springs, radius arms, lever arm hydraulic dampers; Lockheed hydraulic brakes disc front, 7in (17.8cm) drums rear; rack and pinion steering; 6gal (27.3l) fuel tank; racing tyres; wire wheels. DIMENSIONS wheelbase 80in (203.2cm); track 45.25in (114.9cm) front, 44.75in (113.7cm) rear; turning circle 30.1ft (9.2m); ground clearance 4in (10.2cm); length 135.25in (346.1cm); width 53in (134.5cm); height 48in (20.3cm) (121.9cm). PERFORMANCE maximum speed 120mph (192.6kph). PRODUCTION 3.

1964 MGB Sebring racer

Among the 15 or so MGBs campaigned by the Abingdon Competitions Department under Stuart Turner, three were GTs, of which two were special lightweights. Open ones were raced and rallied with and without hardtops and among engine permutations were three or five-bearing blocks and cross-flow or aluminium heads. Compression ratios could be as high as 11:1 and tuning experts such as Daniel Richmond at Downton or Morris Engines were consulted. For Le Mans the cars had smoothed-off nose-cones to improve the aerodynamics, not unlike the Dick Jacobs Midgets, and profiled plastic headlamp covers.

In 1964 Paddy Hopkirk and Andrew Hedges averaged 99.9mph (160.4kph) at Le Mans in No. 37 BMO 54B, finishing 19th overall. At Sebring in 1965 an MGB came second in class and at Le Mans in 1965 Hopkirk and Hedges finished 11th, the best an MGB ever achieved there, and second in its class. It was the last MGB to tackle Le Mans, but glory of another sort was still to come in the Marathon de la Route, the 84-hour race at the Nürburgring. Julien Vernaeve and Andrew Hedges, driving MGB No. 47 GRX 307D, won after 5260 miles (8465km) of racing. It was a great testimony to the strength of the MGB and to team manager Peter Browning. Hopkirk and Hedges won their class at Sebring in a newly built MGB GT LBL 591E in 1967.

BODY Roadster, coupe, 2 doors, 2 seats; weight 2030lb (908kg). ENGINE 4 cylinders, in-line; front; 81.32mm x 88.9mm, 1847cc; compr 8.8:1 or more; 130bhp (96.9kW) @ 6500rpm; 70.4bhp (52.5kW)/l. ENGINE STRUCTURE pushrod overhead valve; chain-driven camshaft; cast iron cylinder head and block; Weber twin choke 45DCOE carburettor; SU electric fuel pump; centrifugal and vacuum ignition control; 5-bearing crankshaft (3-bearing for some events); engine rubber-mounted; oil cooler; tubular exhaust manifold. TRANSMISSION rear wheel drive; Borg & Beck sdp clutch; 4-speed close-ratio gearbox, overdrive fitted according to event; single-piece open prop shaft; Salisbury hypoid bevel final drive 3.91:1 (or to suit event); limited slip differential. CHASSIS steel monocoque structure; ifs by coil springs and unequal wishbones; live axle with semi-elliptic springs, anti-roll bar; Armstrong competition adjustable dampers; disc brakes; rack and pinion steering; 23gal (104.6l) fuel tank (or additional tank to standard); racing tyres; wide rimmed aluminium or wire wheels. DIMENSIONS wheelbase 91in (231.1cm); track 49in (22.9cm) (124.5cm); turning circle 30.5ft (9.3m); length 153.75in (390.5cm). PERFORMANCE maximum speed 140mph (224.7kph) in Le Mans trim; 7kg/bhp (9.4kg/kW). PRODUCTION approx 15.

Chronology of MG Records

1946 Oct Gardner Gardner-MG s/c Jabbeke 750cc 3 at max 159.151mph (256.12kph)

1947 Jul Gardner Gardner-MG s/c Jabbeke 500cc 4 at max 118.061mph (190.00kph

1949 Sep Gardner Gardner-MG s/c Jabbeke 500cc 3 at max 154.86mph (249.22kph)

1950 Jul Gardner Gardner-MG s/c Jabbeke 3 50cc 3 at max 121.09mph (1 94.87kph

1951 Aug Gardner Gardner-MG s/c TD Utah 1500cc 6 at max 137.4mph (221.12kph)

1952 Aug Gardner Gardner-MG s/c TD Utah 1500cc 2 at max 189.5mph (304.96kph)

1954 Aug Eyston, Miles EX179 u/s TF Utah 1500cc 8 at max 153.69mph (247.33kph)

1956 Aug Miles, Lockett EX179 u/s Twin Cam Utah 1500cc 16 at 170.1mph (273.8kph)

1957 Aug Ash, Wisdom EX179 u/s A-series Utah 1100cc 3 at 118.13mph (190.llkph)

Aug Ash, P Hill EX179 s/c A-series Utah 1100cc 6 at max 143.47mph 230.89kp]i)

Aug Moss EX181 s/c Twin Cam Utah 1500cc 5 at max 245.64mph (395.3lkph)

1959 Sep Wisdom Ehrman Leavens EX219 s/c A-series Utah 1100 146.9mph (236.5kph)

Sep P Hill EX181 big s/c Twin Cam Utah 2000cc 6 at max 254.91mph (410.23kph).

MG Racing and Rallying

1945 Coupe de la Libération Race, Paris: 2nd.

1946 Argentine 100 Hours Trial: 1st.

Coupe du Salon Race, Paris: 2nd.

New South Wales GP: 1st 2nd and 3rd.

1947 Australian Grand Prix: 1st 2nd and 3rd.

Chappell Cup, Keller Cup and Colmore Trophy Trials: 1st.

Circuit of Ireland Trial: 1st and 2nd.

Highland Three Days Rally: class win.

Hill climbs: Ballinascorney and Sierre/Montana Hillclimbs: FTD.

Lyons Cup Race, France: 5th.

O'Boyle Trophy Race, Curragh, Ireland: 1st 2nd and 3rd.

Prix de la Blècherette Race, Lausanne, Switzerland: 1st 2nd and 3rd.

Rheims Cup Race, France: 6th, 7th and fastest lap.

Ulster Trophy Handicap Race, Ballyclare, N Ireland: 1st and 2nd.

Watkins Glen, New York: 11 MGs in first race, two driven by Sam and Miles Collier.

1948 Australian Grand Prix: 2nd.

Berne Sports Car GP, Switzerland: 1st.

Hill climbs: Enniskerry, Killakee: FTD.

Leinster Trophy Race: 2nd and 3rd.

O'Boyle Trophy Race, Curragh, Ireland: 1st and 2nd.

Vue des Alpes hillclimb, Switzerland: 1st 2nd and 3rd in class.

Watkins Glen GP, USA: 3rd.

1949 Alpine Trial: 1st and 2nd in class and Coupe des Dames.

Berne Sports Car GP, Switzerland: 2nd.

Circuit of Ireland Trial: Team prize.

Fairfield Memorial Trophy Race, S Africa: 2nd.

Le Mans 24-hour race. George Phillips/'Curly' Dryden (TC Midget) disqualified after giving mechanic illegal lift to pits.

Leinster Trophy Race, Ireland: 1st.

Lisbon Rally, Portugal: 3rd in class.

O'Boyle Trophy Race, Curragh, Ireland: 1st 2nd and 3rd.

Spa/Francorchamps 24 Hours Race, Belgium: 2nd in class.

1950 Alpine Trial: class win and Team award.

Blandford Camp: prototype TD Mark II (EX 171) driven by Dick Jacobs beats HRGs.

Champion Trophy Race, Dundrod, N Ireland: 3rd.

Dundrod TF 3-hour race: 3 'FRX' TD. Mark IIs beat HRGs; 1st 2nd and 3rd in class.

Fairfield Memorial Trophy Race, S Africa: 1st and 3rd.

Le Mans 24-hour race: Phillips/Winterbottom (MG TC) 18th overall at 72.82mph (117.19kph), 2nd in class.

Leinster Trophy Race: 2nd and 3rd.

O'Boyle Trophy Race: 1st and 3rd.

Palm Beach 1300 Rally: 1st and 3rd.

Palm Beach Races: 1st and 2nd in class.

Palm Springs Sports Car Race: 2nd and 3rd.

Production Car Race, Silverstone: 2nd 3rd and 4th in class.

RAC Tourist Trophy Race, N Ireland: 1st 2nd and 3rd in class.

Sebring: first race at Hendrick Field, Sebring, Florida. Six hours. Van Driel (TC 1) 1st in class, 5th overall; O'Hare/Milliken (TC 3) 2nd in class, 6th overall; Keith/ Wilder (TC 25) 2nd in class G, 9th overall; Viall/Charlwood (TD 22) 4th in class, 13th overall; Brundage/Cook (TD 32) do not finish.

Silverstone production car race: TD Mark IIs driven by Dick Jacobs, George Phillips and Ted Lund, 2nd 3rd and 4th in class.

Van Riebeck Trophy Race, S Africa: 1st and 3rd.

Westhampton Races, USA: 1st and 3rd.

1951 Alpine Trial: 3rd in class.

Australian Jubilee Grand Prix: 3rd.

British Empire Trophy Race, Isle of Man: 2nd and 3rd in class.

Champion Trophy Race, Dundrod: 3rd.

Circuit of Ireland Trial: 1st.

Daily Express Rally: 1st.

Interlaken Rally, Switzerland: 1st.

Isle of Wight Rally: 1st 2nd and 3rd in class, Team prize and Coupe des Dames.

Lakeland '300' Rally, USA: 1st.

Le Mans 24-hour race: Phillips/Rippon (TD Midget 43) 116mph (186.68kph) on Mulsanne Straight but retire with engine problems.

Leinster Trophy Handicap, Ireland: 3rd and Team prize; Leinster Trophy Scratch Race: 3rd.

Morecambe Rally: 2nd and 2 class wins.

Paris/St Raphael Rally: 3rd (1st in class).

Production Car Race, Silverstone: class win.

RAC Rally of Great Britain: 1st and 3rd in class.

Rallye des Neiges, France: 2nd.

San Diego Road Race, USA: class win.

Scottish Rally: 2 class wins, Team prize.

Sebring 12-hour race: Ash/Van Driel (1390cc MG MkII Special 4) 2nd in class, 6th overall; Thompson/Kinchloe, Hansgen/Pearsall, O'Hare/Allen (TD Midgets 2, 52 and 5) 3rd 4th and 6th in class (8th 10th and 12th overall), Team prize.

1952 Circuit of Ireland Trial: 1st, Team award.

Circuit of Munster Rally, Ireland: 1st.

Daily Express Rally: 1st 2nd and 3rd in class, Team award.

Edenvale Sports Car Race, Canada: 1st 2nd and 3rd.

Falkirk Rally, Scotland: 1st.

Helsingland Sports Car Race, Sweden: 1st.

Johore Production Car Race, Malaya: 1st 2nd 3rd and 4th.

Leinster Trophy Handicap Race, Ireland: 2nd and 3rd.

O'Boyle Trophy Race, Curragh: 3rd.

Production Car Race, Silverstone: class win.

Sebring 12 Hours Pace, USA: Team prize.

Tulip Rally, Holland: 3rd, 1st in class.

Ulster Trophy 1300cc Race, Dundrod: 1st.

1953 Australian Grand Prix: 2nd and 3rd.

British Empire Trophy Pace, Isle of Man: 3rd in class.

Canyon Rally, Canada: 1st.

Daily Express Rally: 1st 2nd and 3rd in two classes, Team award.

Delaware Valley Trophy Race, USA: 2nd and 3rd.

Evian/Mont Blanc Rally: 3rd in class.

Falkirk Rally, Scotland: 1st.

Great American Mountain Rally, USA: 1st, two class wins.

Johore Production Car Race, Malaya: 1st 2nd and 3rd in class.

Morecambe Rally: class win, Team award.

Omaha National Sports Car Race, USA: 2nd and 3rd.

Production Car Race, Silverstone: 1st and 3rd in class.

RAC Coronation Rally: YB saloons (149, 150 and 151). Geoff Holt 2nd in class, 15th overall; Len Shaw 1st in class, 6th overall; G Holt unplaced.

RAC Rally of Great Britain: 1st 2nd and 3rd in class.

RAC Tourist Trophy Race, N Ireland: class win.

Rally to the Midnight Sun, Sweden: class win.

Sam Collier Memorial Trophy Race, USA: Class win.

Sebring 12-hour race: 1389cc MG special (4) of Ash/Ahrens 2nd in class, 13th overall; Shields/McKinsley (TD Midget 53) 3rd in class, 15th overall; Wellenburg/Wonder (TD 42) 4th in class, 16th overall; MG specials Allen/Longworth (44) 5th in class, 19th overall; Paterson/Brundage (14) 7th in class, 28th overall; Leibensperger /Class (61) 36th overall.

Circuit of Ireland Trial: 3rd in class, Team Award and Ladies award.

Tulip Rally, Holland: Class win.

Winter Handicap Race, S Africa: 1st and 2nd.

1954 Australian Grand Prix: 2nd.

Bogota Sports Car Race, Colombia: 1st 2nd and 3rd in class.

British Redex Rally: 1st and 2nd in class.

Circuit of Ireland Trial: 2nd and 3rd and Team award.

False Bay Production Car Race, S Africa: 1st.

Leinster Trophy Handicap Race: 1st.

Mount Druitt 24 Hours Race, Australia: class win.

Production Car Race, Silverstone: class win.

RAC Rally of Great Britain: class win.

Sebring 12-hour race: Ehrman/Allen 5th in class, 11th overall in MG 'Motto' special (54). MG specials driven by Herzog/Lansing (99), Derujinsky/Underwood (100), and Franklin Curtis (59) dnf.

Westhampton Production Car: class win.

December BMC Competitions Dept set up in former development shop, A-block.

1955 Daily Express Production Touring Car Race, Silverstone: Jacobs, Foster and Waller (ZA Magnettes) 1st 2nd and 3rd.

Dundrod TT: Flockhart/Lockett (prototype twin-cam EX182 34) retire with ignition trouble; Lund/Stoop (1500 36) retire; Fairman/Wilson (1500 35) 4th in class.

Le Mans 24-hour race: three EX 182 prototypes. Miles/Lockett (41) 12th, Lund/Waeffler (64) 17th, Jacobs/Flynn (42) crashes near Maison Blanche at 6.39pm, 2½ hours into the race, minutes after Mercedes crash of Pierre Levegh.

MCC Hastings Rally: 1st and Team award.

Comps department moves to hangar B8.

Monte Carlo Rally: 4 ZA Magnette. Holt/Asbury/Brookes (49 'Aramis') 178th; Holt/Collinson/Cave (58 'Athos') 237th; Len Shaw/Brown/ Finnemore (36 'Porthos') 202nd; D'Artagnan' driven by Chambers.

Montlhéry endurance runs: Ken Wharton (MGA) 102.54 miles (165.02km) in an hour; John Gott (MGA) 112.36mph (180.82kph), tyre burst three quarters distance required re-start.

RAC Rally of Great Britain: Pat Moss/Pat Faichney (TF 193) drive through heavy s snow, 3rd in ladies' competition. Holt/Brookes (Magnette 198) 1st in class.

Scottish Rally: 1st 2nd and 3rd in class, Team prize.

Sebring 12-hour race: TF debut. Ash/Black (76) 8th in class, 38th overall; Ryan/Kinne (77) 9th in class, 39th overall.

Settlers' Handicap Race, S Africa: 2nd and 3rd.

Van Riebeck Trophy Race, S Africa: 2nd.

Alpine Trial: 3rd in class, Coupe des Dames.

1956 Autosport Production Sports Car Championship: winner.

Cape Grand Prix, S Africa: 1st.

Circuit of Ireland Trial: Team award.

Ladies' European Rally Championship: winner.

Liège-Rome-Liège Rally: 4 MGAs (1489cc B-series) entered. Milne/Bensted-Smith (15) 7th in class, 14th overall; Burgess/Croft-Pearson (24) retire with sump damage; Mitchell/Anne Hall (38) 26th overall, 2nd in Ladies' class; Gott/ Tooley (75) 6th in class, 13th overall.

Lyons/Charbonnières Rally, France: Coupe des Dames .

Mille Miglia, Brescia: Nancy Mitchell/Pat Faichney (MGA 227) 3rd in class, 74th overall, Mitchell highest placed female 1624km (1009.13 miles) in 15hr 7min 28sec. Scott-Russell/Haig in (MGA 229) 2nd in class.

Monte Carlo Rally: two alloy-panelled MG Magnette ZAs. Grant/ Norman Davis/Cliff Davis (34) retire with electrical problems; Susan Hindermarsh/Nancy Mitchell/Doreen Reece (327) finish 59th overall, 3rd in Ladies' class.

RAC Rally of Great Britain: 1st 2nd 3rd in class.

Rallye des Alpes: Mitchell/ Faichney (MGA 227) 3rd in class, 15th overall, win Ladies' prize; Milne/Johns (MGA 314) 4th in class; Shephard/Williamson (MGA 308) 5th in class.

Scottish Rally: 2nd in class, Ladies Award.

Sebring 12-hours: three-car team white MGAs prepared by BMC Competitions Department at Abingdon. Driven by Americans, Team prize. Kinchloe/Spittler (50) 4th in class, and 19th; Ash/Ehrman (49) 5th in class, 20th overall; Allen/Van Driel (51) 22nd overall.

Tulip Rally: Mitchell/Reece Magnette (147) excluded for non-standard alloy panels.

1957 Ladies' European Rally Championship: winner.

Leinster Trophy Handicap Race: 2nd.

Liège-Rome-Liège Rally: Mitchell/Johns (MGA 5) 9th in class, 16th, win Ladies' prize; Milne/Shepherd (24) retire due to accident; Gott/Tooley (MGA 48) 8th in class, 14th overall. Comps dept moves to B1 building.

Lyons/CharbonnièresRally, France: Coupe des Dames.

Mille Miglia, Italy: 2nd and 3rd in class.

Production Car Race, Silverstone: 1st and 2nd in class.

Sebring 12 Hours Race, USA: 1st and 2nd in class and Team prize.

Settlers' Handicap Race, S Africa: 3rd.

1958 Autosport Three Hours, Snetterton: 3rd.

BRSCC Saloon Car Championship: class win.

Liège-Rome-Liège: Gott/Brookes (MGA Twin Cam 78) 9th overall (first factory Twin Cam entry in international rallying).

RAC Tourist Trophy Race, Goodwood: 3rd in class.

1959 Acropolis Rally: 2nd and 3rd in class.

Autosport Production Sports Car Championship: class win.

Circuit of Ireland Trial: 3rd in class.

Grand Touring Race, Silverstone: 1st 2nd and 3rd in class.

Le Mans 24-hour race: Lund/Escott (MGA Twin Cam 33) retire in 21st hour.

Pietermaritzburg Six Hours Race: 2nd and 3rd on index.

Rand Nine Hours Race, S Africa: 3rd.

Sebring 12-hour race: four Twin Cams entered by Hambro with BMC Canada. Ehrman/Saidel (28) 2nd in class, 27th overall; Parkinson/Dalton (29) 3rd in class, 34th overall; Flaherty/Pickering/Decker (30) 4th in class, 45th overall.

1960 Autosport Production Sports Car Championship: class win.

Autosport Three Hours, Snetterton: 3rd.

Circuit of Ireland Trial: 2nd in class.

Le Mans 24-hour race: Lund/Escott (32) 1st in class, 12th on dist, 91.12mph (146.64kph).

Phoenix Park Handicap Race: 2nd, 3rd.

RAC Tourist Trophy Race, Goodwood: 3rd in class.

Rallye Deutschland: 2nd in class.

Sebring 12 Hours Race, USA: 3rd in class.

1961 Le Mans 24-hour race: Lund/Olthoff (58) retire after 15 laps.

Monte Carlo Rally: Donald and Erle Morley (MGA 1600 Coupe 314) 1st in class, 28th overall; Riley/Hughes (Midget (44) 1st in class, 33rd overall.

RAC Rally: 1st and 2nd in class.

Sebring 12-hour race: Parkinson/Flaherty (MGA 1600 De Luxe coupe 44) 1st in class, 14th overall; Riley/Whitmore (43) 2nd in class, 16th overall.

1962 Bad Homburg Rally, Germany: class win.

Brands Hatch Six Hours Race: class win. *(pictured)*

Dunlop Hillclimb, Singapore: class win.

Liège-Rome-Liège Rally: Gott/Shepherd (MGA 1600 Mark II Coupe 49) retire.

Monte Carlo Rally: two class wins.

RAC Rally: Rev Rupert Jones/David Seigle-Morris (1100 11) only MG retires, broken piston.

Sebring 12-hour race: Ecurie Safety Fast MGA 1600 Mk II De Luxe Coupes. Sears/Hedges (51) 4th in class, 16th overall; Olthoff/Whitmore (52) 6th in class, 20th overall; Flaherty/Parkinson (63) 5th in class, 17th overall.

Tulip Rally: Aaltonen/Palm (MGA 1600 Coupe 11) 1st in class, 15th overall; Gold/Hughes (Midget 59) 3rd in class, 38th overall.

1963 Autosport Production Sports Car Championship: 3rd, 1st in class.

Grand Touring Race, Silverstone: 1st and 2nd in class.

Le Mans 24-hour race: Alan Hutcheson/Paddy Hopkirk (MGB 31) crash at Mulsanne but finish 12th overall at 91.96mph (147.99kph).

Leinster Trophy Race, Ireland: 3rd.

Monte Carlo Rally: Rev Rupert Jones/Phillip Morgan (Midget 158) 1st in class; Raymond Baxter/McMillen (1100 268) 4th in class.

Nürburgring 1000 Km: 2nd and 3rd in class.

Riverside Sports Car Races, USA: six class wins.

Sebring 12-hour race: Jim Parkinson/ Jack Flaherty (MGB 47), Christabel Carlisle/Denise McCluggage (MGB 48) retire.

Tour de France Automobile: Hedges/Sprinzel (MGB 155) retire after accident.

1964 Austrian Alpine Rally: 1st and 2nd in class.

Grand Touring Race, Silverstone: class win.

Le Mans 24-hour race. Hopkirk/Hedges (MGB 37, BMO 541B) 19th overall, 99.95mph (160.85kph).

Monte Carlo Rally: Don and Erle Morley (MGB 83) win GT class, 17th overall.

Nürburgring 1000 Kilometres Race, Germany: 1st and 2nd in class.

RAC Rally: Fitzpatrick/Handley (MGB 177) retire after accident; Mayman/Domleo (MGB 39) retire with clutch problems.

Sebring 12-hour race: Leslie/Dalton (MGB 47) 3rd in class, 17th overall; Adams/Brennan 4th in class, 22nd overall.

Spa-Sofia-Liège Rally: no MGBs finish.

Tour de France Automobile: Hedges/Sprinzel (MGB 153) retire w head gasket failure.

Warwick Farm Sports Car Race, NSW, Australia: 1st.

1965 Austrian Gold Cup Rally: 1st.

Brands Hatch 1000 Miles Race: 1st, and 1st 2nd and 3rd in class.

Bridgehampton 500 Miles Race, USA: 4th, 1st in class.

GT Constructors' Championship: Midget 2nd in class; MGB 3rd in class.

Kingsway Trophy Race, Phoenix Park, Dublin: 1st 2nd 3rd 4th and 5th.

Le Mans 24-hour race: Hopkirk/Hedges (MGB 39) 11th out of 14 finishers, at 98.25mph (158.11kph), 2nd GT 2litre.

Sebring 12-hour race: Brennan/Morrell (MGB 49) 2nd in class, 10th prototype, 25th overall; Pricard/Pease (MGB 48) 6th in class, 10th GT, 32nd overall. Mac/Hedges (Midget 68) 1st in class, 12th GT, 26th overall.

Targa Florio: Hopkirk/Hedges (Midget GT 'Jacobs' coupe 44) 2nd in class, 11th overall.

MG Timeline

Above: Proposed MGA by Frua.

1940 May: Morris Motors renamed the Nuffield Organization.

1941 Nov: Cecil Kimber sacked by Miles Thomas.

1944 Oct 2: Morris Garages acquires Charles Raworth.

1945 Feb 4: Cecil Kimber killed in railway accident.

Apr: Cowley prepared for MG TC Midget, under direction of H A Ryder.

Sep: Nuffield Metal Products formed, incorporating Morris Pressings Branch, set up in 1939.

Sep 17: First production MG TC (Nr 0252, engine Nr 1163) .

Oct: MG TC Midget launched.

1947 Jan 1: RAC horsepower tax changes to annual tax based on engine size. Replaced a year later by flat annual fee of £10.

May: MG Y-series saloon launched.

Aug: Reginald Hanks and Sydney Smith become directors of Morris Motors.

Nov 19: Sir Miles Thomas resigns as chairman of Nuffield Group. Reginald Hanks becomes vice chairman and restructures board. Sydney Smith takes charge of MG at Cowley and 'Pop' Propert at Abingdon.

Feb: Minister of Supply, George Strauss, announces steel allocated only to manufacturers that export over 75% of output.

Oct 27: Earls Court Motor Show. MG YT open 4-seater tourer. Morris Minor MM launched. Austin and Nuffield Organization in exploratory talks, which cease before Jul 1949.

Dec: US-spec MG TC Midget still right-hand drive.

1948 Oct: First post-war Earls Court motor show, YT tourer introduced.

1949 Jan: Wolseley moves from Ward End, Birmingham to Cowley.

May 1: Riley production transferred from Foleshill to Abingdon. Riley factory becomes part of Morris Engines. Wolseley left machining and tractors as Tractor & Transmissions Branch.

Jul 20: Jack Tallow (ex-Riley) takes over from Propert as general manager Abingdon.

Nov 10: first production MG TD (chassis 0252); last TC (Serial Nr 10,001, chassis 10,251).

1950 Jan 18: MG TD Midget introduced.

Apr 15: EX 135 shown at New York Motor Sports Show.

May: Production MG TD Midget Mk II.

Jun: petrol rationing ends.

Dec: 10,000th Abingdon car: Riley 2½ Litre.

1951 Oct 17: MG YB at Earls Court Motor Show.

1952 Mar 31: Nuffield Motors and Austin form British Motor Corporation (BMC); Leonard Lord deputy chairman and MD, Lord Nuffield chairman.

Oct 21: Leonard Lord agrees to Austin-Healey 100. EX 175 rejected.

Nov 3: John Thornley, assistant general manager, takes over as general manager at Abingdon.

Dec: TD Mark II, produced since May 1950 alongside standard TD, gets enamel badges and black and white MG logos instead of brown and cream. On introduction of TF in 1953 all MG badges are black and white.

Dec 17: Lord Nuffield resigns; LP Lord to be chairman.

1953 Jan 8: TF prototype EX 177 presented to BMC's joint board of management.

Mar: Gerald Palmer's twin-cam engine approved for development at Morris Engines.

Apr 4: MG TD-based open and coupe Arnolt Family-and-Sports Car at New York International Motor Sports Show.

Aug 17: Last TD, chassis no TD/29,915.

Sep: First TF; chassis no TF/501.

Oct 21: MG ZA Magnette and TF Midget at Earls Court show.

Dec: AV Oak, technical director at Cowley since January 1943, retires.

1954 Jun: Drawing office set up at Abingdon. Staff transferred from Cowley.

Jun 6: First drawing for DO1062 (MGA).

Jul: MG TF 1500, Midget name dropped. Terry Mitchell starts work on DO1062.

Aug: Syd Enever starts to develop EX 175.

Sep: Last MG TF 1250.

Dec 1: BMC Competitions Department opens at Abingdon under Marcus Chambers.

1955 Apr: MGA launch postponed to June then September. Last MG TF 1500 no TF/10,100.

May 16: First production MGA made.

Jul 16: First production MGA leaves factory.

Sep 22: MGA at Frankfurt Motor show.

Oct 19: MGA at Earls Court show.

Nov: Alec Issigonis returns to BMC Longbridge.

1956 Feb 1: Don Hayter joins MG design Abingdon as senior layout draughtsman.

May 16: 100,000th MG, a left-hand drive MGA 1500 tourer, built at Abingdon.

Sep: ZB Magnette, MGA Coupe introduced.

Oct 12: Production version of MGA, Vanden Plas alloy hardtop.

Oct 17: Earls Court. MGA Coupe and ZB Magnette Manumatic and Varitone.

1957 Jan 31: Donald Healey shows George Harriman Q1 prototype Austin-Healey Sprite to be made at Abingdon in 1958.

Jul 11: EX181 leaves Southampton on Queen Mary for the USA.

Nov: production of Austin-Healey 100/6 BN4 transferred from Longbridge to Abingdon. Donald and Geoffrey Healey become consultants.

1958 May 20: Austin-Healey Sprite Mark 1 launched.

Jun 19: EX214/1 MGB prototype drawing completed by Don Hayter.

Jul 15: MGA Twin Cam (below) launched; 100 journalists drive four at the Fighting Vehicles R&D establishment at Chobham, Surrey. One crashes, one breaks down.

Nov 12: Pininfarina Magnette Mark III, built at Cowley, introduced.

Dec: ZB Magnette discontinued.

1959 Mar: Austin-Healey 3000 launched. Built at Abingdon.

Apr: First issue of *Safety Fast*.

May: production starts of MGA 1600 with 1588cc B-series engine.

Jul 1: Austin-Healey 3000 replaces 100-Six.

Jul 31: MGA 1600 launched. MGA 1500 earned $60 million from the USA.

Aug 26: Morris Mini Minor (ADO15) launched.

1960 Spring: MGA Twin Cam discontinued.

May: Pressed Steel begins first MGB prototype.

Nov: lnnocenti of Italy launches 998cc Spyder based on Austin-Healey Sprite.

Dec: production of Sprite Mk I ends.

1961 May: Austin-Healey Sprite Mark II launched.

Jun: Midget Mark I (GAN 1) launched. First use of plastic, red-backed, chrome octagon on black shield for grille badge. Later adopted for MGB and 1100. MGA 1600 Mark II.

Jul: Mini Cooper with 997cc A-series engine introduced.

Jul 26: MGB designers abandon radical rear suspension.

Sep: Roy Brocklehurst becomes MG project engineer.

Sep 18: introduction of MG Magnette Mark IV.

Sep 30: Marcus Chambers' place taken by Stuart Turner.

1962 Mar: New York Motor Show. 100,000th MGA from Abingdon displayed in metallic gold with cream leather trim and gold painted wire wheels.

Mar 24: First production MGB, left-hand-drive, Iris Blue, GHN3-102. 12 production cars built by May, including right-hand-drive GHN3-101.

Jul: MGA 1600 Mk II discontinued; final chassis no GHN2-109071.

Sep 20: AD023 MGB launched.

Oct 2: MG 1100 (ADO16G) announced.

Oct 17: MG 1100 and MGB shown at Earls Court. MG Midget Mk I and Austin-Healey Sprite Mk II appear with 1098cc engine. Midget GAN2. Triumph Spitfire introduced.

1963 Apr 4: MGB assembled at BMC Australia, Zetland, New South Wales.

Aug 22: Death of Viscount Nuffield GBE, CH, FRS

1964 Jan: John Thornley awarded OBE for services to Air Training Corps.

Jan: Jacques Coune's MGB Berlinette at Brussels Motor Show.

Feb 11: Design work starts on MGB replacement, EX234.

Mar 9: MG Midget Mark II (GAN3).

Oct: 5-bearing B-series engine supersedes 3-bearing in MGB.

Acknowledgments

In 2001 the publishers thanked Kevin Howe, chief executive officer of MG Rover and his staff, in particular Gordon Poynter, Greg Allport, and Kevin Jones for their help in making the first edition possible. Nothing contributes to knowledge of any make so much as driving its cars, and for that the author thanked generations of public relations officials of MG, BMC, and their predecessors and successors for making available test cars of every MG since the 1950s. His thanks also went to MG owners who enabled him to take the wheel of cars made in the 30 years before 1950. Recognition also went, although not without reservations, to salesmen and restorers responsible for the author's ownership of MGs.

MG cars are so well documented that the publishers' quest for accuracy was challenging. We consulted widely to try and get things right and were grateful to all the MG clubs, in particular to Roche Bentley, Richard Monk, and Richard Ladds of the MG Owners' Club for their unstinting help and encouragement, especially their help in obtaining photographs. Robert Gammage and Peter Browning of the MG Car Club were equally heartening to the compilers of yet another MG book, adding to the wealth of material available. The late Roger Stanbury, who was not only an early MG owner and Vintage Register chairman, but also Dove Publishing's solicitor was of great help and support. Roger provided critical comment at every stage of the book's production and invaluable contacts with MG authorities on whose experience the author drew including Darrell Cocup with his files and memorabilia concerning John Thornley and Syd Enever. These included authors Phil Jennings and Roger Barraclough whose Oxford to Abingdon was a definitive work, Mike Hawke whose K3 Dossier was invaluable, Malcolm Green, and Geoff Radford who generously made the Wilson McComb archives available. Among the MG historians whose work was invaluable were David Knowles the post-1945 equivalent of Jennings and Barraclough, Andrew Roberts, Simon Goldsworthy and Stewart Roy who kindly read the finished manuscript with the benefit of their long experience of MG. Among the photographic sources we used were the National Motor Museum at Beaulieu. As with all Dove Publishing books, our thanks were due to our production consultant David Bann, and Andrew Barron for the jacket design.

Bibliography

Combat, Barré Lyndon, William Heinemann London, 1933.

Circuit Dust, Barré Lyndon, John Miles, 1934

Grand Prix, Barré Lyndon, John Miles 1935

Wheelspin, More Wheelspin, Wheelspin Abroad, CAN May

Racing Round the World, Count Giovani Lurani

Motor Racing and Record-Breaking, George Eyston and Barré Lyndon

Safety Last, George Eyston

Fastest on Earth, George Eyston

Magic MPH, Lt Col ATG Gardner, Motor Racing Publications, 1951

The Life of Lord Nuffield, PWS Andrews, E Brunner, Basil Blackwell Oxford, 1955

Maintaining the Breed, John Thornley, Motor Racing Publications London, 1950-1956

The MG Companion, Kenneth Ullyett, Stanley Paul, 1960

Out on a Wing, Sir Miles Thomas, Michael Joseph London, 1964.

The Bullnose and Flatnose Morris, L Jarman & R Barraclough, David & Charles 1965

MG Magnette K3, Wilson McComb, Profile Publications, Leatherhead, Surrey, 1966

MG 18/80, Wilson McComb, Profile Publications, 1966

MG M-type, Wilson McComb, Profile Publications, 1967

The Bullnose Morris Cowley, L Jarman RI Barraclough, Profile Publications 1967

The MG Story, JN Wherry, Chilton Books, 1967

The Magic of MG, Mike Allison, Dalton Watson 1972

MG The T Series, Richard Knudson (ed) Motorcars Unlimited, Savannah, Ga, USA, 1973

MG The Sports Car America Loved First, and MG: The Sports Car. Richard L Knudson Motorcars Unlimited Oneonta NY, 1975

The Bullnose and Flatnose Morris, Lytton Jarman & Robin Barraclough, 1976

An MG Experience, Dick Jacobs, Transport Bookman Publications London, 1976.

The MGA, MGB, and MGC, Graham Robson, Motor Racing Publications London, 1977

MG and Austin Healey Spridgets, Chris Harvey

The MG A, B, & C, Chris Harvey

The Immortal T Series, Chris Harvey, Oxford Illustrated Press, 1977

MG, Wilson McComb, Osprey London 1978 (Story of MG, Dent, 1972 revised)

British Leyland, The truth about the cars, Jeff Daniels, Osprey, 1980

The T-Series MGs, Graham Robson, Motor Racing Publications London, 1980.

The Sprites and Midgets, Eric Dymock, Motor Racing Publications London, 1981

MG Past and Present, AF Rivers Fletcher, Gentry Books, 1981

MGB, Wilson McComb, Osprey London, 1982

MGA, Wilson McComb, Osprey London, 1983

MG File, Martin Buckley (Ed), Classic and Sportscar Bay View Books, Bideford, 1987

The Cars of BMC, Graham Robson, Motor Racing Publications, 1987

The MG Log, Peter Haining (Ed), Souvenir Press London 1988.

The Kimber Centenary Book, Richard L Knudson (Ed) The New England MG T Register Oneonta NY, 1988.

The Magic of the Marque, Mike Allison, Dalton Watson London 1989, (pb 1972, revised)

Early MG, PL Jennings, PL Jennings, Llanbister, 1989.

MG Sports Cars, Autocar archive, various publishers incl Hamlyn, Bay View, 1980s.

Original MG T Series, Anders Ditlev Clausager, Bay View Books Bideford, 1989.

Original MGB, Anders Ditlev Clausager, Bay View Books

Original MGA, Anders Ditlev Clausager, Bay View Books

Original Sprite & Midget, Terry Horler, Bay View Books

Essential MG T-series Anders Ditlev Clausager, Bay View Books, 1989

MG Midget 1961-1979 and other Brooklands Books' compliations

MGB The Complete Story, Brian Laban, Crowood, 1990

MGB, Lindsay Porter, Haynes, 1992

MGB restoration and maintenance, Jim Tyler Osprey 1992

MG Midget and Austin-Healey Sprite restoration, Jim Tyler Osprey 1993

K3 Dossier, Mike Hawke, Magna Press, Leatherhead, 1992

MGB Illustrated History, Jonathan Wood, Lionel Burrell, G T Foulis, Haynes, 1993.

MG Gold Portfolio, Brooklands Books Ltd, Cobham, Surrey

Spriteley Years, John Sprinzel, Patrick Stephens, 1994

MG Road Cars Vol 1, 4 cylinder ohc 1929-1936, Malcolm Green, Magna 1994

The MG Collection, Richard Monk, Vol 1 pre-war models, Patrick Stephens 1994

The MG Collection, Richard Monk, Vol 2 post war models Patrick Stephens 1995

MG Trials cars, Roger F Thomas, Magna, 1995

Project Phoenix, The Birth of the MGF, Ian Adcock, Bloomsbury, 1996

MG Road Cars, Vol 2, 6 cylinder ohc 1931-1936, Malcolm Green, Magna 1997

MG Sports Cars, Malcolm Green, Bramley Books, 1997

Rallying in a works MG, Len Shaw, Magna, 1997

MG The Untold Story, David Knowles, Windrow & Green, 1997

MG Collectibles, Michael Ellman-Brown, Bay View Books, 1997

MG Britain's Favourite Sports Car, Malcolm Green, Haynes, 1998

Auto Architect, Gerald Palmer, Magna, 1998

MG from A to Z, Jonathan Wood, Motor Racing Publications, Croydon CR0 3RY 1998

Oxford to Abingdon, RL Barraclough and PL Jennings, Myrtle Pub., Cwmfrain, 1998

MGs on Patrol, Andrea Green, Magna, Leatherhead, 1999

The Works MGs, Mike Allison & Peter Browning, Haynes, 2000

MG Saloon Cars, Anders Ditlev Clausager

BMC Competitions Department Secrets. Marcus Chambers, Stuart Turner, Peter Browning, Veloce Publishing, 2015

Don Hayter's MGB Story, Veloce 2014.

Among the sources used in research were the author's archive collections of the Swiss annual Automobil Revue/Revue Automobile published by Hallwag, Automobile Year published by Editions J-R Piccard, Autocourse published by Hazleton, and also of The Motor, The Autocar, Autosport, Motor Sport, Classic Car, Classic & Sportscar, The Automobile, Safety Fast and Enjoying MG, Automobile Quarterly, and Veteran & Vintage magazine, to all of whose proprietors motoring historians owe continuing thanks.

The Author

Motherwell-born, trained as an engineer, Eric Dymock joined the road test staff of *The Motor* in 1962 and has spent his lifetime as a motoring journalist and author.

He was Grand Prix Correspondent of *The Guardian* and *The Observer* 1966-1980, motoring correspondent of *Town* (Haymarket) 1966-1968, and *The Observer* 1980-1982. While motoring correspondent of *The Sunday Times* 1982-1995, he won four Jet Media Excellence Awards including the outright title in 1988. Twice chairman of the Fleet Street Motoring Group, twice winner of the Montagu of Beaulieu Trophy presented by the Guild of Motoring Writers (GoMW) 1997 and 2008 for books on Saab and Bentley, he was also runner-up for his definitive work on world champion driver Jim Clark.

In a wide-ranging career, he was GoMW Regional Motoring Journalist of 2003. Eric compiled TV motoring programmes in the 1970s and 1980s for BBC and Thames TV, contributed regularly to *The Times, The Financial Times, The Daily Telegraph, The Scotsman,* classic car magazines, *Scotland on Sunday* 2002-2006 and *The Spectator Business* 2007. Broadcaster on BBC Radio 4, BFBS and World Service, in 2004 the Association of Scottish Motoring Writers (ASMW) honouring Scots who have achieved excellence in the field of motoring, presented him with the Jim Clark Memorial Award and in 2014 with its President's Award.

Author of: *Champion Year* with Jackie Stewart (Pelham, 1970), *The World of Racing Cars* (Hamlyn 1972), *The Guinness Guide to Grand Prix Motor Racing* (Guinness Superlatives, 1980), *Postwar Sports Cars* (Ebury, 1981), *Austin-Healey Sprite, MG Midget, a Collector's Guide* (Motor Racing Publications 1981, Heel Verlag 1991), *BMW, A Celebration*, (Pavilion UK and Orion USA, Heyne Verlag 1990), *Jim Clark, Tribute to a Champion* (Dove Publishing 1997 and Dove Digital 2011) and *Ecurie Ecosse*, (PJ Publishing 2007). Also under Dove Publishing's imprint, *Rover The First Ninety Years*, *Saab Half a Century of Achievement*, which gained the Montagu Award for 1997, *Honda the UK Story* and *High Speed Diary*. Dove's series of Eric Dymock Motor Books included definitive histories of Audi, Renault, Vauxhall, MG, Jaguar, Ford in Britain and Land Rover. *The Complete Bentley,* published in November 2008, won a second Montagu and is now included among Dove Publishing's ebooks, which include two volumes of Sports Car Classics, compilations of the author's features and motoring columns on sporting cars driven over 50 years.

MG Classics 2 is the most comprehensive account of every model under one cover up to 1965 with detailed specifications, performance and prices. It details MG company history, lists racing and record-breaking, is fully illustrated and crammed with reference material. Perfect for MG owners, fans and motoring historians. Authentic, accurate and distinctive, it is lively and highly readable.

Researched by Eric Dymock, MG, and Austin-Healey owner, who road-tested and drove every MG featured, the MG Classics series is a unique source of facts, figures and analysis.

MG Classics 1 is already available, also as a digital source or paperback. MG Classics 3 will be published in summer 2018 to complete the series.

Visit www.dovepublishing.co.uk to read more about Dove books and Eric Dymock.

Made in United States
Orlando, FL
04 November 2026

90201771R00076